T0186332

Grid Resource Management

Toward Virtual and Services Compliant Grid Computing

CHAPMAN & HALL/CRC
Numerical Analysis and Scientific Computing

Aims and scope:
Scientific computing and numerical analysis provide invaluable tools for the sciences and engineering. This series aims to capture new developments and summarize state-of-the-art methods over the whole spectrum of these fields. It will include a broad range of textbooks, monographs and handbooks. Volumes in theory, including discretisation techniques, numerical algorithms, multiscale techniques, parallel and distributed algorithms, as well as applications of these methods in multi-disciplinary fields, are welcome. The inclusion of concrete real-world examples is highly encouraged. This series is meant to appeal to students and researchers in mathematics, engineering and computational science.

Proposals for the series should be submitted to one of the series editors above or directly to:
CRC Press, Taylor & Francis Group
4th, Floor, Albert House
1-4 Singer Street
London EC2A 4BQ
UK

Published Titles

Grid Resource Management: Toward Virtual and Services Compliant Grid Computing
Frédéric Magoulès, Thi-Mai-Huong Nguyen, and Lei Yu

Numerical Linear Approximation in C
Nabih N. Abdelmalek and William A. Malek

Parallel Algorithms
Henri Casanova, Arnaud Legrand, and Yves Robert

Parallel Iterative Algorithms
Jacques M. Bahi, Sylvain Contassot-Vivier, and Raphael Couturier

Grid Resource Management

Toward Virtual and Services Compliant Grid Computing

Frédéric Magoulès

Thi-Mai-Huong Nguyen

Lei Yu

CRC Press
Taylor & Francis Group
Boca Raton   London   New York

CRC Press is an imprint of the
Taylor & Francis Group, an **informa** business

A CHAPMAN & HALL BOOK

CRC Press
Taylor & Francis Group
6000 Broken Sound Parkway NW, Suite 300
Boca Raton, FL 33487-2742

© 2009 by Taylor & Francis Group, LLC
CRC Press is an imprint of Taylor & Francis Group, an Informa business

No claim to original U.S. Government works
Printed in the United States of America on acid-free paper
10 9 8 7 6 5 4 3 2 1

International Standard Book Number-13: 978-1-4200-7404-8 (Hardcover)

Visit the Taylor & Francis Web site at
http://www.taylorandfrancis.com

and the CRC Press Web site at
http://www.crcpress.com

Warranty

Every effort has been made to make this book as complete and as accurate as possible, but no warranty of fitness is implied. The information is provided on an as-is basis. The authors, editor and publisher shall have neither liability nor responsibility to any person or entity with respect to any loss or damages arising from the information contained in this book or from the use of the code published in it.

Preface

Grid technologies have created an explosion of interest in both commercial and academic domains in recent years. The development of the World Wide Web, which started as a technology for scientific collaboration but was later adopted for use by a multitude of industries and businesses, has illustrated the development path of grid computing. Grid computing has emerged as an important research area to address the problem of efficiently using multi-institutional pools of resources. Grid computing systems aim to allow coordinated and collaborative resource sharing and problem solving across several institutions to solve large scientific problems that could not be easily solved within the boundaries of a single institution. Although the concept behind grid computing is not new as the idea of harnessing unused Central Processing Unit (CPU) cycles to make better use of distributed resources is known from the new age of distributed computing, grid technology offers the potential for providing secure access to remote services promoting scientific collaborations in an unprecedented scale.

As there are always applications (e.g., climate model computations, biological applications) whose computational demands exceed even the fastest technologies available, it is desirable to efficiently aggregate distributed resources owned by collaborating parties to enable processing of a single application in a reasonable time scale. The simultaneous advances in hardware technologies and increase in wide area network speeds have made the primary purpose of the grid more feasible, which is to bring together a given amount of distributed computing and storage resources to function as a single, virtual computer. In addition to inter-operability and to security concerns, the goal of grid systems is also to achieve performance levels that are greater than any single resource could deliver alone.

The notion of computational grids first appeared in the early 1990s, proposed as infrastructures for advanced science and engineering. This notion was inspired by the analogy to power grids, which give people access to electricity, where the location of the electric power source is far away and usually completely unimportant to the consumer. The power sources can be of different types, burning coal or gas or using nuclear fuel, and of different capacity. All of these characteristics are completely hidden to the consumers, who experience only the electric power, which they can make use of for commodity equipment like plugs and cables. In the future, computational power is expected to become a purchasable commodity, such as electrical power. This book attempts to give a comprehensive view of architectural issues of grid technology (e.g., security, data management, logging, and aggregation of services) and related technologies.

Chapter 1 gives a general introduction to grid computing that takes its name from an analogy with the electrical power grid. Although brief, this chapter offers a classification of grid usages, grid systems and the evolution of grid computing. The first generation of grid systems which introduced metacomputing environments, such as I-WAY supporting wide-area high-performance computing, have paved the path for the evolution of grid computing to the next generation. The second generation focused on the development of middleware, such as Globus Toolkit, which introduced more inter-operable solutions. The current trend of grid developments is moving towards a more service-oriented approach that exposes the grid protocols using Web services standards (e.g., WSDL, SOAP). This continuing evolution allows grid systems to be built in an inter-operable and flexible way and to be capable of running a wide range of applications.

Chapter 2 presents the concepts and operational issues associated with the concepts of Web services and Service Oriented Architecture (SOA). This chapter provides information on the Web services standards and the underlying technologies used in Web services, including Simple Open Access Protocol (SOAP), Web Service Description Language (WSDL), and Universal Description, Discovery, and Integration (UDDI). We describe the emergence of a family of specifications, such as OGSA/OGSI, WSRF, and WS-Notification, which enforces traditional Web services with features such as state and lifecycle, making them more suitable for managing and sharing resources on the grid environments.

Chapter 3 presents technical and business topics relevant to data management in grid environments. We begin by identifying the challenges that have arisen from scientific applications as the data requirements for these applications increase in both volume and scale, and we follow by discussing data management needs in grid environments. We then overview main grid activities today in data-intensive grid computing including major data grid projects on a worldwide scale. We also present a classification for existing solutions for managing data in grid environments.

Grid and peer-to-peer systems share a common goal: sharing and harnessing resources across various administrative domains. The peer-to-peer paradigm is a successful model that has been proved to achieve scalability in large-scale distributed systems. Chapter 4 presents a general introduction to peer-to-peer (P2P) computing including an overview of the evolution and characteristics of P2P systems. Then, routing algorithms for data lookup in unstructured, structured, and hybrid P2P systems are reviewed. Finally, we present the shortcomings and improvements for data lookup in these systems.

Chapter 5 presents a grid-enabled virtual file system named GRAVY, which enables the inter-operability between heterogeneous file systems in grid environments. GRAVY integrates underlying heterogeneous file systems into a unified location-transparent file system of the grid. This virtual file system provides to applications and users a uniform global view and a uniform access through standard application programming interfaces (API) and interfaces.

Chapter 6 first introduces several scheduling algorithms and strategies for heterogeneous computing systems. There are eleven static heuristics and two types of dynamic heuristics which are presented. Then scheduling problems in a grid environment are discussed. We emphasize that new scheduling algorithms and strategies must be researched to take the characteristic issues of grids into account. Concurrently, grid scheduling algorithms, grid scheduling architectures and several meta-scheduler projects are presented. Service-Oriented Architecture (SOA) is adopted more and more in industry and business domains as a common and effective solution to resolve the grid computing problem and the efficient discovery of grid services is essential for the success of grid computing. Thus the service discovery, resource information and grid scheduling architecture are also presented in details. As a specific case of application scheduling, data-intensive applications scheduling is then introduced in order to achieve efficient scheduling of data-intensive applications on grids. Finally, fault-tolerant technologies are discussed to deal properly with system failures and to ensure the functionality of grid systems.

Chapter 7 first presents workflow management systems and workflow specification languages. Then the concept of grid workflow is defined and two approaches to create grid workflows are explained. Next, we underline that the workflow scheduling and rescheduling problem is the key factor to improve the performance of workflow applications and workflow scheduling algorithms. In order to hide low-level grid access mechanisms and to make even nonexpert users of grids capable of defining and executing workflow applications, some portal technologies are also presented at the end of this chapter.

Chapter 8 introduces notions of semantic technologies such as semantic web, ontologies and semantic grid. Semantic grid is considered as the convergence of semantic web and grid and this integration of semantic technologies can improve the performance of grids in two main aspects: the discovery of available resources and the data integration. Semantic web service enhances the description level of web services such as their capabilities and task achieving character. Thus this integration provides the support in service recognition, service configuration, service comparison and automated composition. Several models of service composition are discussed and automatic service composition is presented to demonstrate a brilliant prospect for the automatic workflow generation.

Chapter 9 presents a framework for dynamic deployment of scientific applications into grid environment. The framework addresses dynamic applications deployment. The local administrator can dynamically make some applications available or unavailable on the grid resource without stopping the execution of the Globus Toolkit Java Web Services container. An application scheduler has been integrated in this framework, which can realize simple job scheduling, selecting the best grid resource to submit jobs for the users. The performance of the framework has been evaluated by several experiments. All the components in the framework are realized in the standard of Web service, so the other meta-schedulers or clients can interact with the components in a standard way.

Chapter 10 first introduces some of the main concepts of grid engineering. We emphasize that the research of grid applications should focus on the computing model and system structure design because of the existing numerous grid middlewares which deal with the security, resource management, information handling and data transfer issues in a grid environment. Then several large scale grid projects are presented to show the generic architecture of large scale grid systems and development experiences. At the end, the concept of grid service programming is introduced. The Java WS core programming and GT4 Security are two important aspects mentioned in this chapter.

Chapter 11 draws some conclusions. First this chapter concludes the major contributions of this book which consist of two main aspects: data management and execution management. For each aspect, a summary is provided to outline the brief works in the book. Then the possible future of the grid is introduced. We believe that grid computing will continue to evolve in both data management and execution management of the grid community. Finally many interesting questions and issues, that deserve further research are pointed out.

List of Tables

List of Figures

Contents

Chapter 1

An overview of grid computing

1.1 Introduction

Grid computing has emerged as an important field, distinguished from conventional distributed computing by its focus on large-scale resource sharing, innovative applications, and, in some cases, high-performance orientation [24]. The fundamental objective of grid computing is to unify distributed computer resources independent of scale, hardware, and software in order to achieve a processing power in unprecedented ways. In the early 1990s, scientific community realized that high-speed networks presented an opportunity for resource sharing. This would allow interpersonal collaboration, distributed data analysis, or access to specialized scientific instrumentation.

The term "grid" was inspired by the analogy to power grids, which give people access to electricity, where the location of the electric power source is far away and usually completely unimportant to the consumer. The power sources can be of different type, burning coal or gas or using nuclear fuel, and of different capacity. All of these characteristics are completely hidden to the consumers, who experience only the electric power, which they can make use of for commodity equipment like plugs and cables.

1.2 Classifying grid usages

Grid technology aims to combine distributed and diverse resources through a set of service interfaces based on common protocols in order to offer computing support for applications. The different types of computing support for applications can be classified into five major groups [22]:

- *Distributed computing*: applications can use grid to aggregate computational resources in order to tackle problems that cannot be solved on a single system. Therefore, the completion time for the execution of an application is significantly reduced. This type of computing support requires the effective scheduling of resource using, the scalability of protocols and algorithms to a large number of nodes, latency-tolerant algorithms as well as a high level of performance. Typical applications that require distributed computing are very large problems, such as simulation of complex physical processes, which need lots of resources like CPU and memory.

- *High-throughput computing*: the grids can be used to harness unused processor cycles in order to perform independent tasks [22]. In that way, a complicated application can be divided into multiple tasks scheduled and managed by the grids. Applications that need to be performed with different parameter configurations are well suited for high-throughput computing. For example, Monte Carlo simulations, molecular simulations of liquid crystal, bio-statistical problems solved with inductive logic programming, etc.

- *On-demand computing*: the grids can provide access to resources that cannot be cost-effectively or conveniently located locally. On-demand computing support raises some challenging issues, including resource location, scheduling, code management, configuration, fault tolerance, security, and payment mechanisms. A meteorological application that can use a dynamically acquired supercomputer to perform a cloud detection algorithm is a representative example of an application requiring on-demand computing.

- *Data intensive computing*: the grids are able to synthesize new information from distributed data repositories, digital libraries and databases to meet short-term requirements for resources of applications. Challenges for data intensive computing support include the scheduling and configuration of complex, high-volume data flows. The experiments in the high energy physics (HEP) field are typical applications that need data intensive computing support.

- *Collaborative computing*: the grids allow applications to enable and enhance human-to-human interactions. This type of application imposes strict requirements on real-time capabilities and implies a wide range of many different interactions that can take place [33]. An example application that may use a collaborative computing infrastructure is multi-conferencing.

1.3 Classifying grid systems

Typically, grid computing systems are classified into computational and data grids. In the computational grid, the focus lies on optimizing execution time of applications that require a great number of computing processing cycles. On the other hand, the data grid provides the solution for large scale data management problems. In [32], a similar taxonomy for grid systems is presented, which proposes a third category, the service grid.

- *Computational grid*: refers to systems that harness machines of an administrative domain in a "cycle-stealing" mode to have higher computational capacity than the capacity of any constituent machine in the system.

- *Data grid*: denotes systems that provide a hardware and software infrastructure for synthesizing new information from data repositories that are distributed in a wide area network.

- *Service grid*: refers to systems that provide services that are not provided by any single local machine. This category is further divided as on demand (aggregate resources to provide new services), collaborative (connect users and applications via a virtual workspace), and multimedia (infrastructure for real-time multimedia applications).

1.4 Definitions

While grid technology has caused an explosion of interest in both the commercial and academic domain, no exact definition of "the grid" has been given. The definition of the grid changes along with the evolution of grid technology. There exists multiple definitions of the grid. The lack of a complete grid definition has already been mentioned in the literature [17], [57], [24], [27]. We examine in this section some main definitions extracted from the grid literature sources to find the most exhaustive definition of the grid.

- *As a hardware or software infrastructure [22]*: This early definition (i.e., in 1998) of the grid reveals the similarities to the power grid analogy: "A computational grid is a hardware or software infrastructure that provides dependable, consistent, pervasive and inexpensive access to high-end computational capabilities".

- *As distributed resources with networked interface [57], [29]*: The above definition has been refined in [57] by dropping the "high-end" attribute and promoting grids for every hardware level and type: "The computing resources transparently available to the user via this networked environment have been called a metacomputer" or in [29]: "A metasystem is a system composed of heterogeneous hosts (both parallel processors and conventional architectures), possibly controlled by separate organizational entities, and connected by an irregular interconnection network".

- *As a unique and very powerful supercomputer [27]*: "Users will be presented the illusion of a single, very powerful computer, rather than a collection of disparate machines. [...] Further, boundaries between computers will be invisible, as will the location of data and the failure of processors".

- *As a system of coordinated resources delivering qualities of service [57]*: A grid is a system that "coordinates resources that are not subject to centralized control using standard, open, general-purpose protocols and interfaces to deliver non-trivial qualities of service (QoS)".

- *As a hardware or software infrastructure among virtual organizations [26]*: The author complements the above definition by defining the grid as "A hardware and software infrastructure that provides dependable, consistent, and pervasive access to resources to enable sharing of computational resources, utility computing, autonomic computing, collaboration among virtual organizations, and distributed data processing, among others".

- *As a virtual organization [24]*: The focus lies in the notion of *virtual organization (VO)* because the resource sharing involves not only the data file exchange but also direct access to computers, softwares, data and other resources. In the context of large projects, companies and scientific institutes have to collaborate from different sites in order to pool their databases, knowledge bases, simulation or modeling tools, etc. These resources need to be controlled with the agreement on the sharing conditions, security constraints, etc. between the providers and the consumers of resources. The agreement on these conditions among different institutions forms a virtual organization. Its goal is to share data resources, material means, scientific tools, etc. in order to reduce significantly the conception costs. Specifically, the author emphasizes in [24] that: "The real and specific problem that underlies the grid concept is coordinated resource sharing and problem solving in dynamic, multi-institutional virtual organizations".

- *As a virtual computer formed by a networked set of heterogeneous machines [32]*: "A distributed network computing (NC) system is a virtual computer formed by a networked set of heterogeneous machines that agree to share their local resources with each other. A grid is a very large scale, generalized distributed NC system that can scale to Internet-size environments with machines distributed across multiple organizations and administrative domains".

- *As an infrastructure composed of diverse resources in dynamic and distributed VO [23]*: The author invokes that "Grid technologies and infrastructure support the sharing and coordinated use of diverse resources in dynamic, distributed virtual organizations - that is, the creation, from geographically distributed components operated by distinct organizations with differing policies, of virtual computing systems that are sufficiently integrated to deliver the desired QoS".

- *As an approach enabling a shared infrastructure including knowledge resources [37]*: The author observes that grid computing promotes an approach to conducting collaborations between the scientific and business community: "We define the grid approach, or paradigm, that represents a general concept and idea to promote a vision for sophisticated international scientific and business-oriented collaborations".

As time goes by, the definition of the grid becomes more and more general in order to include multiple capabilities expected from this technology. According to the list of definitions extracted from literature that are identified previously, a grid can be defined as:

DEFINITION 1.1 *A hardware and software infrastructure that provides transparent, dependable, pervasive and consistent access to large-scale distributed resources owned and shared by multiple administrative organizations in order to deliver support for a wide range of applications with the desired qualities of service. These applications can perform either high throughput computing, on-demand computing, data intensive computing, or collaborative computing.*

1.5   Evolution of grid computing

The notion of grid computing had already been explored in the very early days of computer science as shown in the Figure 1.1. In 1969, the vision of

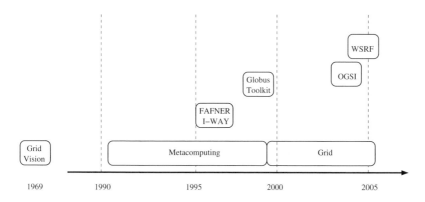

FIGURE 1.1: General technological evolution.

a grid infrastructure was introduced in [31]: *"We will probably see the spread of computer utilities, which, like present electric and telephone utilities, will service individual homes and offices across the country"*.

This vision of wide area distributed computing has become more realistic with the creation of the Internet in the early days of 1990s. The popularity of the wide area network and the availability of inexpensive commodity components have changed the way in which applications are designed. For example, a climatologist may develop his codes, initially on a vector computer to be performed on parallel Multiple Instruction Multiple Data (MIMD) machines. Although these different codes could be run on different machines, they are still considered as a part of the same application. The emergence of a new wave of applications requiring a variety of heterogeneous resources that are not available on a single machine has led to the development of what is know as *metacomputing*. In [34], the authors describe the concept of metacomputing as: *"The metacomputer is, simply put, a collection of computers held together by state-of-the-art technology and "balanced" so that, to the individual user, it looks and acts like a single computer. The constituent parts of the resulting metacomputer could be housed locally, or distributed between buildings, even continents"*. These early metacomputing systems initiated the evolution of grid technology. In [33], the authors summarize the evolution of the grid into three different generations:

- *First generation*: was marked by early metacomputing environments, such as FAFNER [3] and I-WAY [20].

- *Second generation*: was represented by the development of core grid technologies: grid resource management (e.g., Globus, Legion); resource brokers and schedulers (e.g., Condor, PBS); grid portals (e.g., Grid-Sphere); and complete integrated systems (e.g., UNICORE, Cactus).

- *Third generation*: saw the convergence between grid computing and Web services technologies (e.g., OGSI, WSRF).

The next three sections present a brief summary of the key technologies in each stage of grid evolution.

1.5.1 First generation: early metacomputing environments

In the early of 1990s, the first generation efforts were marked by the emergence of metacomputing projects, which aimed to link supercomputing sites to provide access to computational resources. Two representative projects of the first generation are *FAFNER* [3] and *I-WAY* [20], which can be considered as the pioneers of grid computing. FAFNER (Factoring via Network-Enabled Recursion) was created through a consortium to factor RSA 130 using a numerical technique called Number Field Sieve. I-WAY (The Information Wide Area Year) was an experimental high performance network that connected several high performance computers spread over seventeen universities and research centers using mainly ATM technology.

Some differences between these projects are: (i) while FAFNER focused on one specific application (i.e., RSA 130 factorization), I-WAY could execute different applications, mainly high performance applications; (ii) while FAFNER was able to use almost any kind of machine with more than 4MB of memory, I-WAY was supposed to run on high-performance computers with a high bandwidth and low latency network.

Despite these differences, both had to overcome a number of similar obstacles, including communications, resource management, and the manipulation of remote data, to be able to work efficiently and effectively. Both projects also inspire the development of some grid systems. FAFNER was the precursor of projects such as *SETI@home* (The Search for Extraterrestrial Intelligence at Home) [13] and Distributed.Net [12]. I-WAY was the predecessor of the *Globus* [209] and the *Legion* [28] projects.

1.5.1.1   FAFNER

The RSA public key encryption algorithm, which is widely used in security technologies, such as Secure Sockets Layer (SSL) is based on the premise that large numbers are extremely difficult to factorize, particularly those with hundreds of digits. In 1991, RSA Data Security Inc. initiated the RSA Factoring Challenge with the aim to provide a test-bed for factoring implementations and create the largest collection of factoring results from many different experts worldwide. In 1995, FAFNER project was set up by Bellcore Labs., Syracuse University and Co-Operating Systems to allow any computer with more than 4MB of memory to contribute to the experiment via the Web.

Concretely, FAFNER used a new factoring method called Number Field Sieve (NFS) for RSA 130 factorization via computational web servers. A web interface form in HTML for NFS was created. Contributors could invoke CGI (Common Gateway Interface) scripts written in Perl on the web server to perform the factoring through this form. FAFNER is basically a collection of server-side factoring efforts, including Perl scripts, HTML pages, project documentation, software distribution, user registration, distribution of sieving tasks, etc. The CGI scripts do not perform the factoring task themselves; they provide interactive registration, task assignment and information services to clients that perform the actual work. The FAFNER project initiated the development of a wave of web based metacomputing projects (e.g., SETI@home [13] and Distributed.Net [12]).

1.5.1.2   I-WAY

I-WAY, which was developed as an experimental demonstration project for Supercomputing 1995[1] in San Diago is generally considered as the first modern

---

[1]http://www.supercomp.org

grid because this project strongly influenced the subsequent grid computing activities. In fact, one of the researchers leading the I-WAY project was Ian Foster who described later in [21] the close link between Globus Toolkit, which is currently the heart of many grid projects, with metacomputing.

The I-WAY experiment was conceived in early 1995 with the aim to link various supercomputing centers through high performance networks in order to provide a metacomputing environment for high computational scientific applications. The I-WAY's initial objective was to integrate distributed resources using existing high bandwidth networks. Specifically, the resources, including virtual environments, datasets, computers, and scientific instruments that resided across seventeen different U.S. sites, were interconnected by ten ATM networks of varying bandwidths and protocols, using different routing and switching technologies.

The I-WAY consisted of a number of point-of-presence (I-POP) servers, which act as gateways to I-WAY. These I-POP servers were connected by the Internet or ATM networks and accessed through a standard software environment called I-Soft. The I-Soft software was designed as an infrastructure comprising of a number of services, including scheduling, security (authentication and auditing), parallel programming support (process creation and communication) and a distributed file system (using AFS, the Andrew File System). It should be noted that the software developed as part of the I-WAY project (i.e., I-Soft toolkit) formed the basis of the Globus toolkit, which provides a foundation for today's grid software.

1.5.2   Second generation: core grid technologies

The I-WAY project paved the path for the evolution of the grid to the second generation, which focused on the development of middleware to support large scale data access and computation.

DEFINITION 1.2   *Middleware is the layer of software residing between the operating system and applications, providing a variety of services required by an application to function correctly.*

The function of middleware in distributed environments is to mediate interaction between the application and the distributed resources. In a grid environment, middleware continues its role as a means for achieving the primary objective of the grid, which is to provide resources in a simple and transparent way. Grid middleware is designed to hide the heterogeneous nature of resources in order to provide users and applications with a homogeneous and seamless environment.

Some of the technologies that are focused on the second generation grid technologies are the development of grid resource management, resource brokers and schedulers, grid portals, and complete integrated systems. In the next sections, we focus on the evolution of these grid software systems.

1.5.2.1 Grid resource management

The two most representative projects that focus on the development of a grid resource management system are Globus and Legion.

Globus [21] The Globus project is a U.S. research effort initiated by the Argone National Laboratory, University of Southern California's Information Sciences Institute, and University of Chicago with the goal to provide a software infrastructure that enables applications to handle distributed heterogeneous computing resources as a single virtual machine. The most important result of the Globus project is the Globus toolkit (GT) [4]. The GT, a *de-facto* standard in grid computing, is an open source software that focuses on libraries and high-level services rather than end-user applications. It is designed in a modular way with a collection of basic components and services required for building computational grids, such as security, resource location, resource management, and communications. As the components and services are distinct and have well-defined interfaces (APIs), developers of specific tools or applications can exploit them to meet their own particular needs. Specifically, the GT supports the following:

- Grid Security Infrastructure (GSI)

- GridFTP

- Globus Resource Allocation Manager (GRAM)

- Metacomputing Directory Service (MDS-2)

- Global Access to Secondary Storage (GASS)

- Data catalogue and replica management

- Advanced Resource Reservation and Allocation (GARA)

Globus is constructed as a layered architecture in which high-level global services are built upon essential low-level core local services. This architecture is composed of four layers under the application layer [24]. The Figure 1.2 depicts this architecture together with its relationship with the Internet protocol architecture.

Resource and *Connectivity* are the central layers that are responsible for the sharing of individual resources. The protocols of these layers are designed to be implemented on top of the *Fabric* layer, and to be used to build several global services and specific application behavior in the *Collective* layer. The *Fabric* layer is composed of a set of protocols, application interfaces and toolkits to enable the development of services and components to access resources, such as computers, storage resources, and network. The *Collective* layer deals with the coordinated use of multiple resources.

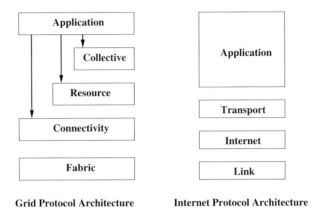

Grid Protocol Architecture   Internet Protocol Architecture

FIGURE 1.2: The layered grid architecture and its relationship to the Internet protocol architecture [24].

Globus arose from the I-WAY project and has evolved a lot from its initial version (GT1) toward a grid architecture based on service-oriented approach (GT4).

Legion [30]   The Legion project developed at the University of Virginia aims to provide a *grid global operating system*, which provides a virtual machine interface layered over the grid. The main objective of the project is to build a global virtual computer, which transparently handles all the complexity of the interaction with the resources of underlying distributed systems (e.g., scheduling on processors, data transfer, communication and synchronization). The focus is to give the users the illusion that they are working on a single computer, with access to all kinds of data and physical resources. Users can create shared virtual work spaces to collaborate research and exchange information and they can authenticate from any machine which has installed Legion middleware to have access on these work spaces as well as secure data transmission when required. Architecturally, Legion is an open system, which aims to encourage third party development of new or updated applications, runtime library implementations, and core components.

The Legion middleware design is based on an object-oriented approach: all of its components (e.g., data resources, hardware, software, computation) are represented as *Legion objects*. It is possible for users to run applications written in multiple languages since Legion supports inter-operability between objects written in multiple languages.

The Legion project began in late 1993 and released its first software version in November 1997. In August 1998, Applied Metacomputing was founded to commercialize the technology derived from Legion. In June 2001, Applied Metacomputing was reformed as Avaki Corporation [1].

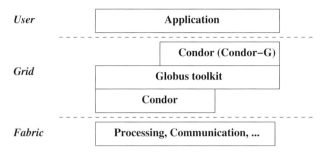

FIGURE 1.3: Condor in conjunction with Globus technologies in grid middleware, which lies between the user's environment and the actual fabric (resources) [36].

1.5.2.2 Grid resource brokers and schedulers

During the second generation, we saw the tremendous growth of grid resource brokers and scheduler systems. The primary objective of these systems is to couple commodity machines in order to achieve the equivalent power of supercomputers with a significantly less expensive cost. A wide variety of powerful grid resource brokers and scheduler systems, such as Condor, PBS, Maui scheduler, LSF, and SGE, spread throughout academia and business.

Condor [2] The Condor project, developed at the University of Wisconsin-Madison, introduces the Condor High Throughput Computing System, which is often referred to simply as Condor and Condor-G.

- *The Condor High Throughput Computing System [35]* is a specialized workload management system for executing compute intensive jobs on a variety of platform environments (i.e., Unix and Windows). Condor provides a job management mechanism, scheduling policy, priority scheme, resource monitoring, and resource management. The key feature of Condor is the ability to scavenge and manage wasted CPU power from idle desktop workstations across an entire organization. Workstations are dynamically placed in a resource pool whenever they become idle and removed from the resource pool when they get busy. Condor is responsible for allocating a machine from the resource pool for the execution of jobs and monitoring the activity on all the participating computing resources.

- *Condor-G [25]* is the technological combination of the Globus and Condor projects, which aims to enable the utilization of large collections of resources spanning across multiple domains. The Globus contribution is the use of protocols for secure inter-domain communications and standardized access to a variety of remote batch systems. Condor contributes the user concerns of job submission, job allocation, error recovery and creation of a user-friendly environment. Condor technology provides solutions for both the frontend and backend of a middleware as shown in the Figure 1.3. Condor-G offers an interface for reliable job submission and management for the whole system. The Condor High Throughput Computing system can be used as the fabric management service for one or more sites. The Globus toolkit can be used as the bridge interfacing between them.

Portable Batch System (PBS) [9]   The PBS project is a flexible batch queuing and workload management system originally developed by Veridian Systems for NASA. The primary purpose of PBS is to provide controls for initiating and scheduling the execution of batch jobs. PBS operates on a variety of networked, multi-platform UNIX environments, from heterogeneous clusters of workstations to massive parallel systems. PBS supports both interactive and batch mode, and provides a friendly graphical interface for job submission, tracking, and administrative purposes.

PBS is based on the client-server model. The main components are *pbs_server* server process, which manages high-level batch objects such as queues and jobs, and *pbs_mom* server process, which is responsible for job execution. The *pbs_server* receives submitted jobs from users in the form of a script and schedules the job for later execution by a *pbs_mom* process.

PBS consists of several built-in schedulers, each of which can be customized for specific requirements. The default scheduler in PBS maximizes the CPU utilization by applying the first-in-first-out (FIFO) method. It loops through the queued job list and starts any job that fits in the available resources. However, this effectively prevents large jobs from ever starting since the required resources are unlikely ever to be available. To allow large jobs to start, this scheduler implements a "starving jobs" mechanism, which defines circumstances under which starving jobs can be launched (e.g., first in the job queue, waiting time is longer than some predefined time). However, this method may not work under certain circumstances (e.g., the scheduler would halt starting of new jobs until starving jobs can be started). In this context, the Maui scheduler was adopted as a plug-in scheduler for the PBS system.

Maui scheduler [16]   The Maui scheduler, developed principally by David Jackson for the Maui High Performance Computer Center, is an advanced batch job scheduler with a large feature set, well suited for high performance computing (HPC) platforms. The key to the Maui scheduling design is its wall-time based reservation system, which allows sites to control exactly when, how, and by whom resources are used. The jobs are queued and managed based upon their priority, which is specified from several configurable parameters.

Maui uses a two-phase scheduling algorithm. During the first phase, the scheduler starts jobs with highest priority and then makes a reservation in the future for the next high priority job. In the second phase, the Maui scheduler uses the backfill mechanism to ensure that large jobs (i.e., starving jobs) will be executed at a certain moment. It attempts to find lower priority jobs that will fit into time gaps in the reservation system. This gives large jobs a guaranteed start time, while providing a quick turn around for small jobs. In this way, the resource utilization is optimized and job response time is minimized. Maui uses the fair-share technique when making scheduling decisions based on job history.

Load Sharing Facility (LSF) [8]   LSF is a commercial resource manager for clusters from Platform Computing Corporation[2]. It is currently the most widely used commercial job management system. LSF focuses on the management of a broad range of job types such as batch, parallel, distributed, and interactive. The key features of LSF include system supports for automatic and manual checkpoints, migrations, automatic job dependencies and job re-schedulings.

LSF supports numerous scheduling algorithms, such as first-come-first-served, fair-share, backfill. It can also interface with external schedulers (e.g., Maui) that complement features of the resource manager and enable sophisticated scheduling.

Sun Grid Engine (SGE) [10]   SGE is a popular job management system supported by Sun Microsystems. It supports distributed resource management and software/hardware optimization in heterogeneous networked environments.

A user submits a job to the SGE, together with the requirement profile, user identification, and a priority number for the job. The requirement profile contains attributes associated with the job, such as memory requirements, operating system required, available software licenses, etc. Then, jobs are kept waiting in a holding area until resources become available for execution. Based on the requirement profile, SGE assigns the job to an appropriate queue associated with a server node on which the job will be executed. SGE maintains load balancing by starting new jobs on the least loaded queue to spread workload among available servers.

1.5.2.3   Grid portals

One of the areas of grid application that is focused on at this time is the development of gateways and grid portals, which are a web-based single point

---

[2]http://www.platform.com

of entry to a grid and its implemented services. With the widespread development of the Internet, scientists expect to expose their data and applications through portals. The grid portal provides a user-friendly web page interface allowing grid applications users to perform operations on the grid and access grid resources specific to a particular domain of interest.

Currently, there are various technologies and toolkits that can be used for grid portal development. According to [38], grid portals can be classified into non portlet-based and portlet-based.

- *Non portlet-based portal*: is a grid portal that is based on a typical three-layers architecture. The first layer is the user layer, which aims to provide the user-friendly interface for users. The user layer is responsible for displaying the portal content; it can be a web browser or other desktop tools. The second layer is the grid service layer, including authentication service, job management service, information service, file service, security service. The authentication service allows the portal to authenticate users. Once authenticated, users can use other services to access resources of the system (e.g., job management service for submitting jobs on a remote machine, information service for monitoring jobs submitted, and viewing results). The second layer receives HTTP requests from the first layer and interacts with the third layer for performing the grid operations on relevant grid resources and retrieving the executed result from grid resources. The third layer is a backend resource layer, which consists of computation, data and application resources.

- *Portlet-based portal*: includes a collection of portlets. A portlet is a web component that generates fragments - pieces of markup (e.g., HTML, XML) adhering to certain specifications (e.g., JSR-168 [7], WSRP [11]). Portlets improve the modular flexibility of developing grid portals as they are pluggable and can be aggregated to form a complete web page conforming to user needs.

In this section, we briefly describe two typical grid portals for each type: Grid Portal Development Kit (GPDK) and GridSphere.

Grid Portal Development Kit (GPDK) The GPDK is a widely used toolkit for building non portlet-based portals. The GPDK is a collaboration between NCSA, SDSC and NASA IPG, which aims to provide generic user and application portal capabilities. It facilitates the development of grid portals and allows various portals to inter-operate by supporting a common set of components and utilities for accessing various grid services using the Globus infrastructure. A GPDK provides a portal development framework for the development and deployment of application-specific portals and a collection of grid service beans for remote job submission, file staging, and querying of information services from a single, secure gateway.

The portal architecture is based on a three-tier model, where a client browser securely communicates to a web server over a secure connection (via https). The web server is capable of accessing various grid services using the Globus infrastructure. The Globus toolkit provides mechanisms for securely submitting jobs to a Globus gatekeeper, querying for hardware/software information using LDAP, and a secure PKI infrastructure using GSI.

The GPDK is based on the Model-View-Controller paradigm and makes use of commodity technologies including the open source servlet container Tomcat, Java Server Pages (JSP), Java Beans, the web server Apache and the Java Commodity Grid (CoG) toolkit.

GridSphere [6]   GridSphere is a typical portlet-based portal. The Grid-Sphere portal framework is developed as a key part of the European project GridLab [98]. It provides an open-source portlet-based web portal and enables developers to quickly develop and package third-party portlet web applications that can be run and administered within the GridSphere portlet container. Two key features of the GridSphere framework are: (i) allowing administrators and individual users to dynamically configure the content based on their requirements, and (ii) supporting grid-specific portlets and APIs for grid-enabled portal development. However, the main disadvantage of the current version of GridSphere (i.e., GridSphere 2.1) is that it does not support WSRP specification.

1.5.2.4   Integrated systems

The widespread emergence of grid middleware has motivated the development of various international projects that integrate these components into coherent systems.

Cactus [14]   Cactus is an open source problem-solving environment designed for scientists and engineers. It supports multiple platforms and has a modular structure, which easily enables the parallel computation across different architectures and collaborative code development between different groups. Cactus originated in the academic research community, where it was developed and used over many years by a large international collaboration of physicists and computational scientists.

Cactus' architecture consists of modules (thorns) which plug into core code (flesh) containing the APIs and infrastructure to adhere the thorns together. The Cactus Computational Toolkit is a group of thorns providing general computational infrastructure for many different applications.

UNiform Interface to COmputing REsources (UNICORE) [15]
UNICORE was originally conceived in 1997 by a consortium of German universities, research laboratories, and software companies. It is funded in part

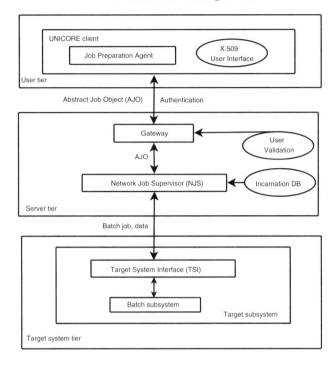

FIGURE 1.4: The UNICORE architecture.

by the German Ministry for Education and Research (BMBF). UNICORE attempts to enable supercomputer centers to provide their users with a seamless, secure, and Internet-based access to the heterogeneous computing resources at the geographically distributed centers.

The UNICORE architecture is based on the three-tier model including user, server and target system tiers as shown in the Figure 1.4. The user tier consists of the graphical user interface - UNICORE client.

A UNICORE job is created using the Job Preparation Agent (JPA), which allows the user to specify the actions to be performed, the resources needed and the system on which the job will be executed. The UNICORE client generates an Abstract Job Object (AJO) from this job description and connects to a Gateway, which authenticates the client before managing the submitted UNICORE jobs. The Gateway transfers the AJO to the Network Job Supervisor (NJS), which translates the abstract job represented by the AJO into a target system specific batch job using the Incarnation Database (IDB).

UNICORE's communication endpoint is the Target System Interface (TSI), which is a daemon executing on the target system. Its role is to interface with the local operating system and the local native batch subsystem.

1.5.3 Third generation: service oriented approach

The core middleware for the grid developed in the second generation provides the basic inter-operability that was crucial to enable large-scale computation and resource sharing. The emergence of service oriented architecture promotes the reusability of existing components and information resources to assemble these components in a flexible manner. The third generation focuses on the adoption of this service oriented model in development of grid applications. The key idea of this solution is to allow the flexible assembly of grid resources by exposing the functionality through standard interfaces with agreed interpretation. This facilitates the easy deployment of grid systems on all scales. Extending Web services to enable transient and stateful behaviors, the Global Grid Forum defined the Open Grid Services Architecture (OGSA) based on Web services protocols, such as WSDL, UDDI and SOAP, described in Chapter 2. OGSA was then combined with grid protocols to define the Open Grid Service Infrastructure (OGSI), which provides a uniform architecture for the development and deployment of grids and grid applications. The creation of Web Services Resource Framework (WSRF), which evolves and refactors OGSI to enable the inter-operability between grid resources using new Web services standards, completes the convergence between web and grid service architecture.

1.6 Concluding remarks

This chapter has presented the concepts of grid computing, which is analogous to the power grid in the way that computing resources will be provided in the same way as gas and electricity are provided to us now. Grid computing has moved from metacomputing environments, such as I-WAY which supports wide-area high-performance computing to grid middlewares and Globus toolkit, which introduces more inter-operable solutions. The current trend of grid development is moving toward a more service oriented approach that exposes the grid protocols using Web services standards (e.g., WSDL, SOAP). This continuing evolution allows grid systems to be built in an inter-operable and flexible way, capable of running a wide range of applications.

References

[1] Avaki. Available online at: http://www.avaki.com (Accessed August 31st, 2007).

[2] Condor. Available online at: http://www.cs.wisc.edu/condor (Accessed August 31st, 2007).

[3] FAFNER. Available online at: http://www.npac.syr.edu/factoring. html (Accessed August 31st, 2007).

[4] Globus toolkit. Available online at: http://www.globus.org/toolkit (Accessed August 31st, 2007).

[5] GridLab. Available online at: http://www.gridlab.org (Accessed August 31st, 2007).

[6] GridSphere. Available online at: http://www.gridsphere.org (Accessed August 31st, 2007).

[7] Introduction to JSR-168. Available online at: http://developers.sun. com/prodtech/portalserver/reference/techart/jsr168/ (Accessed August 31st, 2007).

[8] Platform Computing Inc. Platform LSF. Available online at: http: //www.platform.com/Products/Platform.LSF.Family/ (Accessed August 31st, 2007).

[9] Portable Batch System. Available online at: http://www.openpbs.org (Accessed August 31st, 2007).

[10] Sun Grid Engine. Available online at: http://gridengine.sunsource.net (Accessed August 31st, 2007).

[11] WSRP: Web services for remote portlets. Available online at: http:// www.oasisopen.org/committees/tc_home.php?wg_abbrev=wsrp (Accessed August 31st, 2007).

[12] Distributed.Net, 2004. Available online at: http://www.distributed.net (Accessed August 31st, 2007).

[13] SETI@home: The search for extraterrestrial intelligence at home, 2004. Available online at: http://setiathome.ssl.berkeley.edu (Accessed August 31st, 2007).

[14] G. Allen, T. Dramlitsch, I. Foster, N. T. Karonis, M. Ripeanu, E. Seidel, and B. Toonen. Supporting efficient execution in heterogeneous

distributed computing environments with Cactus and Globus. In *Supercomputing '01: Proceedings of the 2001 ACM/IEEE conference on Supercomputing (CDROM)*, pages 52–52, New York, NY, USA, 2001. ACM Press.

[15] J. Almond and D. Snelling. UNICORE: Uniform access to supercomputing as an element of electronic commerce. *Future Generation Computer Systems*, 15(5–6):539–548, 1999.

[16] B. Bode, D. M. Halstead, R. Kendall, Z. Lei, and D. Jackson. The portable batch scheduler and the Maui scheduler on linux clusters. In *ALS'00: Proceedings of the 4th conference on 4th Annual Linux Showcase and Conference*, pages 27–27, Berkeley, CA, USA, 2000. USENIX Association.

[17] M. L. Bote-Lorenzo, Y. A. Dimitriadis, and E. Gómez-Sánchez. Grid characteristics and uses: A grid definition. In *Proceedings of the First European Across Grids Conference*, volume 2970 of *Lecture Notes in Computer Science*, pages 291–298, Santiago de Compostela, Spain, February 2003. Springer.

[18] I. Foster. What is the grid? A three point checklist. *Grid Today*, 1(6), 2002.

[19] I. Foster. Globus Toolkit version 4: Software for service-oriented systems. In *IFIP International Conference on Network and Parallel Computing*, volume 3779 of *Lecture Notes in Computer Science*, pages 2–13. Springer-Verlag, 2005.

[20] I. Foster, J. Geisler, B. Nickless, W. Smith, and S. Tuecke. Software infrastructure for the I-WAY high-performance distributed computing experiment. In *HPDC '96: Proceedings of the 5th IEEE International Symposium on High Performance Distributed Computing*, page 562, Washington, DC, USA, 1996. IEEE Computer Society.

[21] I. Foster and C. Kesselman. Globus: A metacomputing infrastructure toolkit. *The International Journal of Supercomputer Applications and High Performance Computing*, 11(2):115–128, 1997.

[22] I. Foster and C. Kesselman. *The Grid: Blueprint for a New Computing Infrastructure*. Morgan Kaufmann Publishers, San Francisco, CA, USA, July 1998.

[23] I. Foster, C. Kesselman, J. M. Nick, and S. Tuecke. Grid services for distributed system integration. *Computer*, 35(6):37–46, 2002.

[24] I. Foster, C. Kesselman, and S. Tuecke. The anatomy of the grid: Enabling scalable virtual organizations. *International Journal High Performance Supercomputer Applications*, 15(3):200–222, August 2001.

[25] J. Frey, T. Tannenbaum, M. Livny, I. Foster, and S. Tuecke. Condor-G: A computation management agent for multi-institutional grids. *Cluster Computing*, 5(3):237–246, July 2002.

[26] W. Gentzsch. Response to Ian Foster's: What is the grid? *Grid Today*, 1(8), 2002.

[27] A. Grimshaw. What is a grid? *Grid Today*, 1(26), 2002.

[28] A. Grimshaw, A. Ferrari, F. Knabe, and M. Humphrey. Wide-area computing: Resource sharing on a large scale. *IEEE Computer*, 32(5):29–37, may 1999.

[29] A. S. Grimshaw, J. B. Weissman, E. A. West, and E. C. Loyot, Jr. Metasystems: An approach combining parallel processing and heterogeneous distributed computing systems. *Journal of Parallel and Distributed Computing*, 21(3):257–270, 1994.

[30] A. S. Grimshaw, W. A. Wulf, and C. T. L. Team. The legion vision of a worldwide virtual computer. *Communications of the ACM*, 40(1):39–45, jan 1997.

[31] L. Kleinrock. UCLA to build the first station in nationwide computer network, July 1969. Available online at: `http://www.lk.cs.ucla.edu/LK/Bib/REPORT/press.html` (Accessed August 31st, 2007).

[32] K. Krauter, R. Buyya, and M. Maheswaran. A taxonomy and survey of grid resource management systems for distributed computing. *International Journal of Software Practice and Experience*, 32(2):135–164, 2002.

[33] D. D. Roure, M. A. Baker, N. R. Jennings, and N. R. Shadbolt. *Grid Computing: Making the Global Infrastructure a Reality*, chapter The Evolution of the Grid, pages 65–100. John Wiley and Sons Ltd. Publishing, New York, 2003.

[34] L. Smarr and C. E. Catlett. Metacomputing. *Communications of the ACM*, 35(6):44–52, 1992.

[35] T. Tannenbaum, D. Wright, K. Miller, and M. Livny. Condor: A distributed job scheduler. In T. Sterling, editor, *Beowulf Cluster Computing with Linux*. MIT Press, Oct. 2001.

[36] D. Thain, T. Tannenbaum, and M. Livny. Distributed computing in practice: The Condor experience. *Concurrency - Practice and Experience*, 17(2-4):323–356, 2005.

[37] G. von Laszewski and K. Amin. *Grid Middleware*, chapter Middleware for Communications, pages 109–130. John Wiley, 2004. Available online at: `http://www.mcs.anl.gov/~gregor/papers/vonLaszewski--grid-middleware.pdf` (Accessed August 31st, 2007).

[38] X. Yang, M. T. Dove, M. Hayes, M. Calleja, L. He, and P. Murray-Rust. Survey of major tools and technologies for grid-enabled portal development. In *Proceedings of the UK e-Science All Hands Meeting 2006*, Nottingham, UK, September 2006.

Chapter 2

Grid computing and Web services

2.1 Introduction

Today scientific collaborations require more resources (e.g., CPU, storage, networking) than what can be provided by any single institution. Grid computing is a form of distributed computing that aims to harness computational and data resources at geographically dispersed institutions into a larger distributed system that can be utilized by the entire collaboration. Such a global distributed system, which is dedicated to solving common problems, is known as a virtual organization (VO). Each institution in the VO has its own set of usage policies that it would like to enforce on its resources and services. At the same, it has its local requirements about hardware configuration, operating systems, software toolkits, and communication mechanisms. These differing requirements can result in a very heterogeneous character of grid environments. In this context, Services Oriented Architecture (SOA) emerges as a well suited concept to address some of the issues that arise from such a heterogeneous, locally controlled but globally shared system and the interoperability of applications. Moreover, SOA is considered as the key technology to ease the costs of deployment and maintenance of distributed applications that deliver functionality as *services* with the additional emphasis on loose coupling between integrating services. It provides solutions for business-to-business (B2B) integration, business process integration and management, content management, and design collaboration for computer engineering.

By leveraging Web services - an implementation of SOA, grid computing aims to define standard interfaces for business services and generic reusable grid resources. This convergence effort between grid computing and Web services has lead to the new "service paradigm" to support resource integration and management. In this paradigm, all resources on the grid are treated in a uniform way by being provided a common interface for access and management. From this perspective, a workstation cluster is seen as a "compute service", a database containing scientific data as a "data service", a scientific instrument used to measure seismic data (for instance) as a "data capture service", etc. Each service may be remotely configured and interrogated by a user to identify its interface.

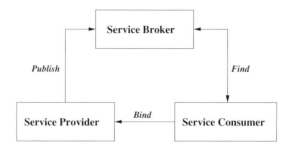

FIGURE 2.1: Web services components can be classified into *service providers*, *service consumers*, and *service brokers*.

2.2 Web services

SOA represents an abstract architectural concept for software development based on loosely coupled components (services) that have been described in a uniform way and that can be discovered and composed. Web services represents one important approach of realizing SOA.

The core idea of a Web service design is simple: a Web service is decoupled from the underlying hardware and software and available to the other services through a well-defined interface. This service can be published, located and invoked by other services over the Internet/intranet. Their functionalities are based on strict standards to enable communication and interactions among services in a simple, easy and seamless manner. Web services model typically involves loosely coupled entities including: *service providers*, *service consumers*, and *service brokers* (see Figure 2.1).

- *Service provider*: is an application that has the ability to perform certain functionality. It makes resources available to service consumers as independent services. A service provider is a self-contained, stateless business function that accepts one or more requests and returns one or more responses through a well-defined, standard interface.

- *Service consumer*: is an application that wants to use the functionality provided by a service. The service consumer sends a message to the provider and requests a certain service.

- *Service broker*: maintains a repository that stores information on the available services and their locations. It is contacted by the service provider, who announces its services and contact information. The service broker is queried by service consumers to obtain the location of a service.

Service providers implement a service and publish it in a service broker or a registry; service consumers locate services in a service registry and then invoke the service. The connection between these entities is loosely coupled offering the maximum decoupling between any two entities. A service consumer does not have to be aware of the implementation of the service provider. This abstraction of the service from the actual implementation offers a variety of advantages to both service providers and service consumers. Service providers can upgrade their internal implementation without impact on their clients. Similarly, service consumers are not forced to adapt the same IT configuration as their service providers. They may choose from several service providers that provide the identical functionality.

2.2.1 Web services characteristics

A typical service exhibits the following defining characteristics [72]:

- *Functional and non-functional*: Services are described in a description language that provides functional and non-functional characteristics. The functional characteristics represent the operational characteristics that define the overall behavior of the service. The non-functional characteristics specify the quality attributes of services, such as authentication, authorization, cost, performance, accuracy, integrity, reliability, scalability, availability, response time, etc.

- *State*: Services could be stateless or stateful. Stateless services can be invoked repeatedly without having to maintain context or state; i.e., an instance of service is stateless if it cannot retain prior events. For example, a travel information service does not keep any memory of what happens to it between requests. In the case of stateful service, it maintains some state between different operation invocations issued by the same or different clients or applications; i.e., it can retain its prior actions. For example, a typical e-commerce service consists of a sequence of stateful interactions involving exchange of messages between partners. The state of a business process needs to be retained in order to undertake a series of interrelated tasks to finish the business process: purchase order, bank transfer, taxation, acknowledgement, shipping notices, etc.

- *Transient-ness*: Services can be transient or non-transient. A transient service instance is one that can be created and destroyed, usually created for specific clients and does not outlive its clients. In contrast to a transient service, a non-transient service or persistent service is designed without the concept of service creation and destruction and outlives its clients.

- *Granularity*: Granularity refers to the scope of functionality provided by a service. The concept of granularity can be applied in two ways: coarse-grained and fine-grained services. Services are called coarse-grained if they provide significant blocks of functionality with a simple invocation. For example, a coarse-grained service might handle the processing of a complete purchase order. By comparison, fine-grained service might handle only one operation in the purchase order process. A fine-grained interface is meant to provide high flexibility for construction of coarse-grained services.

- *Complexity*: Services can vary in function from simple requests to complex systems where the system accesses and combines information from multiple sources. Simple service requests may have complicated realizations. For example, travel plan services are the actual front-end to the complex physical organizational business processes. Typically, a complex service is a coarse-grained service, which involves interactive fine-grained services.

- *Synchronicity*: Services can be distinguished between two programming styles for services: synchronous or Remote Procedure Call (RPC)-style versus asynchronous or message (document)-style. Synchronous services or method-driven services require a tightly coupled model of communication between the client and service provider to maintain the bilateral communication between them. Clients of synchronous services express their request as a method call with an appropriate set of arguments and expect a prompt response containing a return value before continuing execution. On the other hand, asynchronous services or message-driven services allow clients to send an entire document, such as purchase order, rather than a discrete set of parameters. The service accepts the entire document, processes it and may or may not return a result message. Asynchronous services promote a loose coupling between the clients and server because the client that invokes an asynchronous service does not need to wait for a response before it continues with the remainder of its execution. Message driven services are useful where the client does not require (or expect) an immediate response and process-oriented service.

2.2.2 Web services architecture

Web services address the fundamental challenges that distributed computing has provided: providing a uniform way of describing components or services within a network, locating them, and accessing them. The difference between the Web services approach and traditional approaches (e.g., distributed object technologies such as the Object Management Group - Common Object Request Broker Architecture (CORBA), or Microsoft Distributed Component Object Model (DCOM)) lies in the loose coupling aspects of architecture. Instead of building applications that result in tightly integrated collections of objects or components, which are well known and understood at development time, it is more flexible and dynamic to conceive and develop the applications

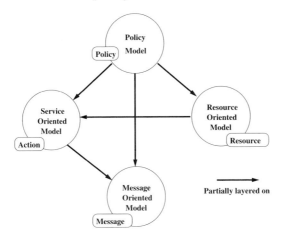

FIGURE 2.2: Meta model of Web services architecture [84].

from loosely coupled services. Another key difference is that Web services architecture is based on standards and technology that are the foundation of the Internet.

There exist various kinds of realizations of SOA proposed by different enterprise-software vendors. Each vendor is trying to define Web services in a slightly different way according to their business and Web services strategies. Therefore, it is a fundamental requirement for inter-operability of higher-level infrastructure services to define a generic Web services architecture in terms of framework and methodology.

2.2.2.1 Generic Web services architecture

A generic Web services architecture aims to provide a consistent way for development of scalable, reliable Web services. There are many architectures and programming models proposed from different vendors like BEA system's WebLogic, IBM's Websphere, Microsoft's .NET Platform, CORBA, Enterprise Edition (J2EE) Enterprise Java Beans which aim to fulfill the goal of a Web services standard. However, these architectures bring with them different assumptions about infrastructure services that are required. Consequently, it is difficult to construct applications from components that are built using different architectures and programming models.

Significant work to address the inter-operability issue has been done through a generic Web services architectural model proposed by the standardization organization World Wide Web Consortium (W3C) [42]. This architecture describes the key concepts and relationships between four models (see Figure 2.2).

- *Message Oriented Model (MOM)*: focuses on messages, message structure (i.e., headers and bodies), message transport (i.e., mechanisms used to deliver messages). There are also additional details to consider, such as the role of policies and how they govern the message level model.

- *Service Oriented Model (SOM)*: builds on the MOM with focusing on aspects of service and action rather than message. The SOM explains the interaction between agent services in using messages in the MOM. It also uses the metadata from the SOA model to document many aspects of services.

- *Resource Oriented Model (ROM)*: focuses on the resource aspects that are relevant to the architecture. Concretely, it focuses on the issues of ownership of resources, policies related to these resources and so on.

- *Policy Oriented Model (POM)*: focuses on constraints on the behavior of agents and services. This model describes the policies imposing constraints to the behavior of agents, people or organizations that attempt to access the resources. Policies may be modeled to represent security concerns, quality of service concerns, management concerns and application concerns.

2.2.2.2 Web services architecture stack

The fact that Web services architecture is composed of several interrelated technologies implies implementation of a stack of specific, complementary standards [84]. The conceptual levels of the architectural stacks provided by [84], [72], [67] are similar in many aspects. Figure 2.3 shows a typical Web services architecture stack. It can be seen that the upper layers build upon the capabilities provided by the lower layers. Likewise, the vertical towers represent requirements that must be addressed at every level of the stack. The text on the left represents standard technologies that apply at that layer of the stack [67].

The core technologies that play a critical role in this architecture stack are XML, SOAP, WSDL and UDDI. These technologies which are widely accepted and implemented uniformly as open standards will be presented in Section 2.3.

2.3   Web services protocols and technology

The World Wide Web Consortium (W3C), which has managed the evolution of the technologies related to Web services (i.e., SOAP, WSDL), defines Web services as: *"A software system designed to support inter-operable machine-to-machine interaction over a network. It has an interface described*

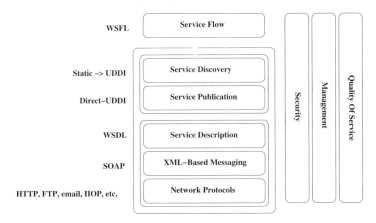

FIGURE 2.3: Web services architecture stack [67].

in a machine-processable format (specifically WSDL). Other systems interact with the Web service in a manner prescribed by its description using SOAP messages, typically conveyed using HTTP with XML serialization in conjunction with other web-related standards." [84].

The Web services approach is based on a maturing set of widely accepted standards. This widespread acceptance enforces the inter-operability between clients and services. Therefore WSDL, XML, SOAP, and UDDI that provide a mechanism for clients to dynamically find other Web services across the network are known as core Web services technology. This section describes these technologies that constitute the Web services standards.

2.3.1 WSDL, UDDI

2.3.1.1 Web Service Discovery Language (WSDL)

WSDL [48] was initially proposed by IBM and Microsoft by merging Microsoft's SOAP Contract Language (SCL) and Service Description Language (SDL), together with IBM's Network Accessible Service Specification Language (NASSL). The first version 1.0 of WSDL was released in September 2000. It has been submitted to the W3C for consideration as a recommendation [85].

WSDL is an XML-based language, which defines the interface of a service. WSDL is similar to "Interactive Data Language" (IDL), which is used to characterize CORBA interfaces. WSDL allows services to be defined in terms of functional characteristics, what actions or functions the service performs, message structures, sequences of message exchanges. In other words, a WSDL document describes what the service can do, where it resides, and how the service can be invoked. It provides a standard view of services provided to clients. Hence, it enforces inter-operability across the various programming paradigms, such as CORBA, J2EE, and .NET.

A WSDL document contains two parts: *abstract definitions* and *concrete descriptions*. The Figure 2.4 outlines the structure and the major parts of a WSDL document, together with their relationships. The abstract section defines operations of a service and its SOAP messages in a language and platform-independent way. In contrast, the concrete descriptions define the bindings of the abstract interface to concrete message formats, protocols (e.g., SOAP, HTTP, and MIME) and endpoint addresses through which the service can be invoked [51].

A WSDL document describes a service as a set of abstract items called *ports* or *endpoints*. A WSDL document also defines abstractly the actions performed by a Web service as *operations* and the data transmitted to these actions as *messages*. A collection of related operations is known as a *Port-Type*. A PortType constitutes the collection of actions offered by the service. The operations and messages are described abstractly and then tied to a concrete transport protocol and data encoding scheme through a *binding*. A binding specifies the transport protocol and message format specifications for a particular PortType. A port is defined by associating a network address with a binding. If a client locates a WSDL document and finds the binding and network address for each port, it can call the service's operations according to the specified protocol and message format. The following paragraph summarizes WSDL document elements from [48], [77].

- *Message Parts*: are a flexible mechanism for describing the logical abstract content of a message. A binding may reference the name of a part in order to specify binding-specific information about the part.

- *Message*: defines an abstract message that can serve as the input or output of an operation. Messages consist of one or more part elements, which can be of different types. For example, each message part can be associated with either an element (when using document style) or a type (when using RPC style).

- *Operation*: is an abstract description of an action supported by the service.

- *PortType*: defines a set of operations performed by the Web services, also known as interface. Each operation contains a set of input, output, and fault messages. The order of these elements defines the message exchange pattern supported by the given operation.

- *Binding*: defines message format (i.e., data encoding) and protocol details (i.e., messaging protocol, underlying communication protocol) for operations and messages defined by a particular PortType. The number of bindings for a particular PortType can be extensible.

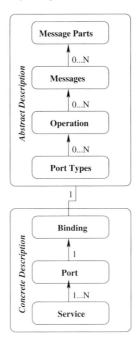

FIGURE 2.4: A WSDL document structure [51].

- *Port*: specifies a single address for a binding, also known as endpoint. In other words, a port element contains endpoint data, including physical address and protocol information.

- *Service*: defines a collection of ports or endpoints by grouping a set of related ports together.

With WSDL, a service can be defined, described and discovered irrespective of its implementation details. In other words, the implementation for a Web service can be done in any language, platform, object model, or messaging system. The application needs to provide a common format to encode and decode messages to and from any number of proprietary integration solutions such as CORBA, COM, EJB, JMS, COBOL [84].

2.3.1.2 Universal Detection and Discovery Interface (UDDI)

UDDI [40] is a widely acknowledged specification for definition of the way in which Web services are published and discovered across the network. The first version of UDDI 1.0 specification was developed by Ariba, IBM, and Microsoft in September 2000. The current version of the UDDI 3.0.2 specification was released in October 2004 [39].

If WSDL describes the service, UDDI stores the description of services itself. UDDI allows a service provider to register information about the services they offer so that other clients can find them. It provides an inter-operable, foundational infrastructure based on a common set of industry standards, including HTTP, XML, XML Schema, and SOAP, UDDI for a Web services-based software environment for both publicly available services and services exposed only internally within an organization [39].

The core component of UDDI is a XML based business registration, which consists of *white pages* including address and contact identifiers, *yellow pages* including categorization based on standards and *green pages* containing technical information about the service.

2.3.2 Web services encoding and transport

Data flow exchanged between programs needs to be converted into a format that is understood by sender and receiver. Common formats for Web services are based on XML encoding methods, including SOAP and XML-RPC, which is an early implementation of the SOAP standard. The process of creating an XML representation of application internal data is called *serialization*. The inverse process of generating application internal structures from XML is called *de-serialization*.

Serialized data is transferred over the network by a specific transport protocol. It should be noted that data transport is independent of the encoding. Web services may be built on top of nearly any transport protocol. The most popular transport protocols of Web services are network protocols, such as Hypertext Transfer Protocol (HTTP) [206], Simple Mail Transfer Protocol (SMTP) [75], or File Transfer Protocol (FTP) [225]. The Web services message exchange is independent of the chosen transport layer. This "transport-neutral" property makes Web services an inter-operable messaging architecture.

2.3.2.1 Extended Markup Languages - Remote Procedure Call (XML-RPC)

XML-RPC is an XML-based standard for making simple remote calls across the network [86] using HTTP as transport and XML as encoding. It emerged in early 1998 as the ancestor of the SOAP protocol. XML-RPC is an extremely lightweight mechanism that can be used as a part of a Web service architecture. It provides the necessary functionality to specify data types and parameters, and to invoke remote procedures in a platform-neutral way.

Data structures   XML-RPC defines eight data types, including six primitive types (see Table 2.1) and two complex types (i.e., Structures and Arrays).

- *Structures*: identify a value with a string-typed key. Structures can be nested: the value tags can enclose sub-substructures or arrays.

Table 2.1: XML-RPC primitive types.

Type	Value	Examples
int or i4	32-bit integers between 2.147.483.648 and 2.147.483.647	`<int>42</int>` `<i4>42</i4>`
double	64-bit floating-point numbers	`<double>3.1415</double>` `<double>-1.4165</double>`
boolean	true (1) or false (0)	`<boolean>0</boolean>` `<boolean>1</boolean>`
string	ASCII text, though many implementations support Unicode	`<string>Paris</string>` `<string>hello!</string>`
dataTime.iso8601	Dates in ISO8601 format: $CCYYMMDDटHH:MM:SS$	`<dateTime.iso8601>` `19040101T05:24:54` `</dateTime.iso8601>`
base64	Binary information encoded as Base 64, as defined in RFC 2045	`<base64>` `SGVsbG8sIFdvcmxkIQ==` `</base64>`

```
<struct>
    <member>
        <name>Key</name>
        <value>Value</value>
    </member>
    <member>
        <name>Key</name>
        <value>Value</value>
    </member>
</struct>
```

- *Arrays*: contain a list of value. The values do not need to be of homogeneous type. Within the value tags, any of the primitive types are allowed. Arrays can also contain sub-arrays and structures.

```
<array>
    <data>
        <value>...</value>
        <value>...</value>
    </data>
</array>
```

Request/response structure XML-RPC defines the format of method calls and responses. The XML message body contains a tag to indicate the method name to be invoked and the parameter list. The server returns a value back to the client when a successful call is completed. The sequence diagram of a XML-RPC request/response cycle is shown in Figure 2.5. Here's an example of an XML-RPC request [86]:

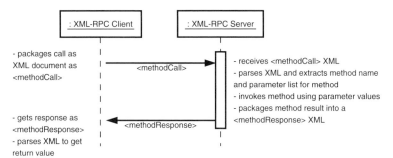

FIGURE 2.5: XML-RPC request structure.

```
POST /RPC2 HTTP/1.0
User-Agent: Frontier/5.1.2 (WinNT)
Host: betty.userland.com
Content-Type: text/xml
Content-length: 181

<?xml version="1.0"?>
<methodCall>
      <methodName>examples.getStateName</methodName>
      <params>
         <param>
            <value><i4>41</i4></value>
         </param>
      </params>
</methodCall>
```

An XML-RPC message is sent through an HTTP POST request. The body of the message is in XML. The message causes a procedure to be executed on the server. Parameters for this procedure are included in the XML message body. The value returned by the procedure is also encoded in XML. In this example, the message request containing `methodCall` is sent to the server to retrieve the state name of a region. The string value 41 is supplied as the argument for the `examples.getStateName` method, which is invoked on the server side. The XML-RPC response returned by the server contains the `methodResponse` and a state name reply of type string, e.g., `<string>Paris</string>`.

2.3.2.2 Simple Object Access Protocol (SOAP)

SOAP was initially created by DevelopMentor, Microsoft, and Userland Software. Microsoft solicited industry feedback on the SOAP 0.9 specification in September 1999. The most recent version of SOAP 1.2 [83] was standardized by W3C.

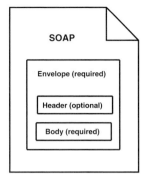

FIGURE 2.6: The structure of a SOAP document.

SOAP is also an XML-based, platform-independent protocol providing a simple and relatively lightweight mechanism for exchanging structured and typed information between services over the network. The lightweight feature of SOAP protocol is explained by two fundamental properties: (i) sending and receiving HTTP (or other) transport protocol packets, and (ii) processing XML messages. SOAP is designed with the aim to reduce the cost and complexity of integrating applications that are built on different operating systems, programming environments, or object model frameworks. For example, applications developed using distributed communication technologies such as CORBA, DCOM, Java/RMI or any other application-to-application communication protocols have a symmetrical requirement for the communication between them. In other words, both ends of the communication link would need to be implemented under the same distributed object model and would require the deployment of libraries developed in common [72]. SOAP offers a standard, extensible, composable framework for packaging and exchanging XML messages [84] within heterogeneous platforms over the network. SOAP may use different protocols such as HTTP, SMTP, FTP, JMS, etc. to transport messages, locate the remote system and initiate communications. However, SOAP's natural transport protocol is HTTP.

SOAP defines an extensible enveloping mechanism for structuring the message exchange between services. A SOAP message is an XML document that consists of three distinct elements: an *envelope*, a *header*, and a *body* (see Figure 2.6).

- *SOAP envelope*: is the root of SOAP message, which wraps the entire message containing an optional header element and a mandatory body element. It defines a framework for describing what is in a message and how to process it. All elements of the SOAP envelope are defined by a W3C XML Schema (XSD) [72].

- *SOAP header*: is a generic mechanism for adding extensible features to SOAP, such as security and routing information.

- *SOAP body*: contains the payload (i.e., application-specific XML data) intended for the receiver who will process it, in addition to the optional fault element for reporting errors occurred during the messages processing. The body must be contained within the envelope, and must follow any headers defined for the message.

Apart from defining an envelope for describing the content of a message and details for how to process it, three other parts are specified within SOAP protocol: a set of data encoding rules, a usage convention, and SOAP binding framework. Data encoding rules define how instances of data types, which are defined by an application, are expressed in a SOAP message, such as float, integer, arrays, etc. Usage conventions define how a SOAP message can execute across the network by specifying a SOAP communication model: Remote Procedure Call (RPC) or document-style communication. The SOAP binding framework specifies the transport protocol through which SOAP messages are exchanged to an application.

2.3.2.3 SOAP versus XML-RPC

While XML-RPC performs remote procedure calls at only a simple level, SOAP reaches for more complex features and has more capabilities. SOAP overcomes the limitations of XML-RPC about the limited type system by providing more robust data typing mechanisms based upon XML Schema [66] (even allowing the creation of custom data types). Since the most remarkable feature of XML-RPC is its simplicity, it is easier to use XML-RPC compared to SOAP despite its limited capabilities, while SOAP provides more utilities and it is less natural to use [76].

SOAP involves significantly more overhead but adds much more information about what is being sent. If complex user-defined data types and the ability to have each message defined how it should be processed are needed, then SOAP is a better solution. In contrast, if standard data types and simple method calls are enough then XML-RPC makes applications faster and easier to develop and maintain.

2.3.3 Emerging standards

The primary standards on which Web services are built are XML, SOAP, UDDI, and WSDL. They constitute a basic building block for Web services architecture and address the inter-operability between services across the network. They ensure that a consumer and a provider of service can communicate to each other irrespective of the location and implementation details of service. However, for a Web services-based SOA to become a mainstream IT practice, other standards may be considered: "higher-level" standards need to be developed and adopted. This is especially true in the areas of Web services security and Web services management. Various standards organizations, such as the World Wide Web Consortium (W3C) and OASIS, have drafted standards in these areas that promise to gain universal acceptance. Two emerging standards of special interest are WS-Security and WS-BPEL [71].

2.3.3.1 Web Services Security (WS-Security)

The WS-Security specification was originally published in April 2002 by IBM, Microsoft, and VeriSign. In March 2004, WS-Security [70] was released as OASIS standard, which proposes a standard set of SOAP extensions that provides message integrity and confidentiality for building secure Web services.

WS-Security uses XML Signature [46] to ensure message integrity, which means that a SOAP message is not modified while traveling from a client to its final destination. Similarly, WS-Security uses XML Encryption [68] to provide message confidentiality, which means that a SOAP message is seen only by intended recipients. Specifically, WS-Security defines how to use different types of security tokens, which are a collection of claims made by the sender of a SOAP message, for authentication and authorization purposes. For example, a sender is authenticated by combining a security token with a digital signature, which is used as proof that the sender is indeed associated with the security token. Additionally, WS-Security also provides a general-purpose mechanism for associating security tokens with messages, and describes how to encode binary security tokens.

The specification is designed to be extensible (i.e., support multiple security token formats) and no specific type of security token is required. For example, a client might define one security token for sender identities and provide another security token for their particular business certifications.

WS-Security is flexible and is designed to be used with a wide variety of security models and encryption technologies, such as Public Key Infrastructure (PKI), Kerberos, and Secure Socket Layer (SSL)/Transport Layer Security (TLS).

2.3.3.2 Web Services Business Process Execution Language (WS-BPEL)

Web services provide an evolutionary approach for building distributed applications that facilitate loosely coupled integration and resilience to change. As services are designed to be loosely coupled and to exist independently from each other, they can be combined (i.e., composed) and reused with maximum flexibility. Complex business processes, such as handling of a purchase order, require involving multiple steps performed in a specific sequence that lead to the invocation and interaction of multiple services. For this business process to work properly, the service invocations and interactions need to be coordinated (i.e., service coordination is also known as "orchestration").

Service coordination allows creating a new and more complex service instance that other applications can use. A complex application can be composed from various granular services coordinated in different manners, such as correlated asynchronous service invocation, long running processes or orchestrating autonomous services. WS-BPEL 2.0 [45], which is also identified as BPELWS, BPEL4WS, or simply BPEL, was approved as an official OASIS standard in 2007 for composition and coordination of Web services. WS-BPEL uses WSDL to describe the Web services that participate in a process and how the services interact with each other.

2.4   Grid services

Web services and grid computing are key technologies in distributed systems that attracted a lot of interest in recent years. Web services, which emerged in the year 2000, address the problem of application integration by proposing an architecture based on loosely coupled distributed components and widely accepted protocols and data formats, such as HTTP, WSDL, SOAP.

The concept of grid computing was introduced in 1995 and aims to provide the computational power and data management infrastructure necessary to support the collaboration of people, together with data, tools and computational resources [56]. The primary goal of grid computing is to address computationally hard and data-intensive problems in science and engineering. Different from Web services, which are based on strict standards to enable communication and interaction among applications, the majority of grid systems have been built based on either ad-hoc public components or proprietary technologies [78]. The fact that the interfaces for individual grids were not standardized has led to inter-operability problems of grid systems in large-scale since the communication between grids is usually based on vendor-specific protocols. In current practice, there exist various public and commercial grid middlewares, which have been successful in their niche areas (e.g., Globus). However, due to the lack of a dominant standard among them, these solutions have limited potential as the basis for future-generation grids, which will need to be highly scalable and inter-operable to meet the needs of global enterprises.

Even though starting from different perspectives, there is considerable overlay between the goals of Web services and grid computing initiatives. In fact, both Web services and grid computing deal with service concepts and both architectures have the same underlying design principles provided by SOA. The rapid advances in Web services technology and standards have provided an evolutionary path from the ad-hoc architecture of current grids to the standardized and service-oriented grid of the future [78].

Significant progress has been made in converging these two initiatives in key areas where the efforts overlap with each other. The Globus alliance [41], a broad, open development group for grid computing, offers mechanisms

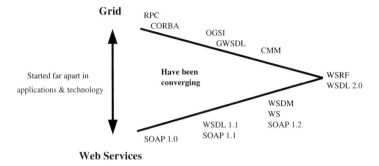

FIGURE 2.7: Convergence of Web services and grid services [69].

for developing grids through Web services. Open Grid Services Architecture (OGSA) [53] is introduced in version 3.0 of Globus Toolkit as an implementation of Open Grid Services Infrastructure (OGSI) proposed by Global Grid Forum (GGF). OGSA refines the architecture of grid computing to address SOA principles and adopts the Web services approach to enhance the capabilities of the grid environment. The technologies used to implement the required services and their specific characteristics are not specified in OGSA. The technical details of how to build the services are defined in OGSI through a set of extensions and specializations to the Web services technology for grid deployment, as required by OGSA. OGSI defines the mechanisms for creating, managing and exchanging information among entities called *grid services*, which are "a Web service that provides a set of well-defined interfaces and that follows specific conventions. The interfaces address discovery, dynamic, service creation, lifetime management, notification, and manageability" [57]. The details about the grid services specification can be found in [81]. The relationship between grid services and Web services is given in detail in [65].

Lately, the collaboration between Web services and the grid computing community [49] has resulted in the important specification Web Services Resource Framework (WSRF) [44], which essentially retains all the functional capabilities present in OGSI, and at the same time builds on broadly adopted Web services concepts. The Figure 2.7 shows the convergence between Web services and grid computing. In the sections that follow we will describe OGSI and WSRF specification.

2.4.1 Open Grid Services Infrastructure (OGSI)

Typically, Web services implementations are stateless. However, for a lot of applications, it is desirable to be able to maintain a state. Especially in grid computing, the state of a resource or service is often important and may need to persist across transactions. For example, an online reservation system must maintain a state about previous reservations made, availability of seats, etc. It is possible by using standard Web services to manage and manipulate stateful (i.e. maintain state information between message calls) services using *ad hoc* methods (e.g., extra characters placed in URLs or extra arguments to functions) across multiple interactions. In other words, the message exchanges that Web services implement are usually intended to enable access to stateful resources.

However, the management of stateful resources acted upon by the Web service implementation is not explicit in the interface definition. This approach requires client applications being aware of the existence of an associated stateful resource type for Web services and it does not address the core issue of state management in general. The lack of standard conventions, which is critical for inter-operability within loosely coupled service-oriented platforms, leads to increased integration cost between Web services that deal with stateful resources in different ways. Therefore, it is desirable to define Web services conventions to solve the fundamental problem of state management in the Web services architecture. A general solution is needed to enable the discovery of, introspection on, and interaction with stateful resources in standard and inter-operable ways. Most important, such an approach improves the robustness of design time selection of services during application assembly and runtime binding to specific resource instances [54].

In this context, OGSI specification version 1.0 [82], released in July 2003 by GGF OGSI Working Group as a base infrastructure on which OGSA is built, specifies a set of extensions of Web services technology to enable stateful Web services. It defines the standard interfaces, behaviors, and core semantics of a grid service. In this specification, the grid service is referred to as a service that conforms to a set of conventions of WSDL [48] and XML Schema [66] relating to its interface definitions and behaviors.

OGSI enhances Web services technology by introducing the concept of stateful and transient services with standard mechanisms for declaring and inspecting state data of a service instance; asynchronous notification of service state change; representing and managing collections of service instances through referenceable handles; lifecycle management of service instances; and common handling of service invocation faults [54].

- *Stateful and transient services*: The fact that grid services can be created as stateful and transient service is considered as one of the most important improvements with regard to Web services.

  Service Data is the OGSI approach to stateful Web services and provides a standard way for service consumers to query and access the state data from a service instance. Since plain Web services allow operations to be included only in the WSDL interface, Service Data can be considered as an extension to the WSDL that allows not only operations but also attributes to be included in the WSDL interface. It is important to note that Service Data is much more than simple attributes; it can be any type of data (e.g., fundamental types, classes, arrays). In general, the

Service Data included in a service will fall into one of two categories: (i) *state information*, which provides information on the current state of the service, such as operation results, intermediate results, runtime information, and (ii) *service metadata*, which is information on the service itself, such as system data, supported interfaces, cost of using the service.

Factory/instance is the OGSI approach to overcome the non-transient limitations of Web services. Since plain Web services are non-transient (i.e., persistent) their lifetime is bound to the Web services container. The fact that all clients work on the same instance of a Web service implies that the information the Web service is maintaining (e.g., computation results) for a specific client may be accessed (and potentially messed up) by any other client. Factories may create transient instances with limited lifetime, which will be destroyed when the client has any use for them. It should be noted that a grid service could be persistent, just like a normal Web service. Choosing between persistent grid services or factory/instance grid services depends entirely on the requirements of the client application.

- *Asynchronous notification*: OGSI provides a mechanism for asynchronous notification of state change. A grid service can be configured to be a notification source by implementing NotificationSource PortType, and certain clients to be notification sinks (or subscribers) by implementing NotificationSink PortType. This allows subscribed clients (sinks) to be notified of changes that occur in a service (source).

- *Collection of service instances*: OGSI enables a number of services to be aggregated together to act as a service group for easier maintenance and management. A grid service can define its relationship with other member services in the group. Services can join or leave a service group.

- *References*: OGSI uses Grid Service Handles (GSH) to name and manage grid service instances. The GSH is returned when a new grid service instance is created as a unique identity. GSH is a global standard URI name for a service instance, which must be registered with the HandleResolver for appropriate invocation. In fact, GSH does not contain sufficient information to allow a client to communicate directly with the service instance, but it may resolve to a Grid Service Reference (GSR). GSR provides the means for communicating with the service instance (e.g., what methods it has, what kind of messages it accepts/receives). The GSR can be a WSDL document for the service instance, which specifies the handle-specific bindings to facilitate service invocation. The client has to hold at least one GSR for interactions with the identified service instance. The grid service instance needs to implement the HandleResolver PortType, that maps a GSH to one or more GSRs, to manage and translate between GSH into GSRs.

Table 2.2: Summary of base PortTypes defined in OGSI specification [82].

PortType Name	Description
GridService	encapsulates the root behavior of the service model, must be implemented by all grid services
HandleResolver	creating an instance of a Grid service returns a Grid Service Handle (GSH). This GSH is mapped to a reference Grid Service Reference (GSR), which then has enough information to enable client communication with the actual instance of a grid resource via a grid service. This interface provides the functionality to map a GSH to a GSR.
NotificationSource	allows clients to subscribe to notification messages
NotificationSubscription	defines the relationship between a single NotificationSource and NotificationSink pair
NotificationSink	defines a single operation for delivering a notification message to the service instance that implements the operation
Factory	is standard operation for creation of grid service instances
ServiceGroup	allows clients to maintain groups of services
ServiceGroupEntry	defines the relationship between a grid service instance and its membership within a ServiceGroup
ServiceGroupRegistration	allows grid services to be added and removed from a ServiceGroup

- *Lifecycle management*: gives a client the ability to create and destroy a service instance according to its requirements.

The OGSI 1.0 specification defines the following PortType (i.e., interfaces) that should be implemented by a grid service. Table 2.2 provides the name of OGSI PortType, its operation and description of such interfaces.

In order to be qualified as a grid service instance, a Web service instance must implement a port whose type is, or is derived from GridService PortType, which specifies the functions that can be called on the service. The service may optionally implement other PortType from the standard OGSI family as listed in the previous table along with any application-specific PortTypes, as required. The GridService PortType has the following operations [82]:

- *findServiceData*: allows a client to discover more information about the service's state, execution environment and additional semantic details that are not available in the GSR. In general, this type of reflection is an important property for services. It can be used to allow the clients a standard way to learn more about the service they will use. The exact way this information is conveyed is through ServiceData elements associated with the service.

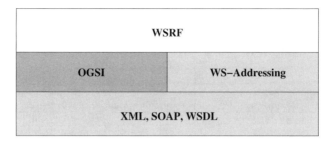

FIGURE 2.8: The relationship between WSRF, OGSI and Web services technologies.

- *setServiceData*: allows the client to modify the value of the Service Data element. This modification implies changing the corresponding state in the underlying service instance.

- *requestTerminationAfter*: allows the client to specify the termination time after which the service instance has to terminate itself.

- *requestTerminationBefore*: allows the client to specify the termination time before which the service instance has to terminate itself.

- *destroy*: explicitly instructs the destruction of the service instance.

In summary, the OGSI specification is an attempt to provide an environment where users can access grid resources through grid services, which are defined as an extension of Web services.

The importance of a concept that addresses the stateful Web services has been recognized by major Web services communities. However, OGSI was not widely accepted by these communities, and concerns were raised about the relationship between this specification and existing Web services specifications as: "too much stuff in one specification", "not working well with existing Web services and XML tooling", "too object-oriented", and "introduction of forthcoming WSDL 2.0 capability as unsupported extensions to WSDL 1.1" [54]. As a result, there has been collaboration among Globus alliance [41], IBM, and HP, towards aligning OGSI functions with emerging advances on Web services technology. This effort produced the concept of Web Services Resources Framework (WSRF) [44] in January 2004. This specification supersedes OGSI and completes Grid and Web services convergence. Figure 2.8 shows the relationship between the Web services technologies and the OGSI and WSRF specifications.

2.4.2 Web Services Resource Framework (WSRF)

The WSRF specification is an evolution of OGSI 1.0, which aims to address the needs of grid services in conjunction with the evolution of Web services.

Table 2.3: WS-Resource Framework specifications summary [50].

Specification Name	Description
WS-ResourceLifetime	Mechanisms for WS-Resource destruction, including message exchanges that allow a requestor to destroy a WS-Resource, either immediately or by using a time-based scheduled resource termination mechanism.
WS-ResourceProperties	Definition of a WS-Resource, and mechanisms for retrieving, changing, and deleting WS-Resource properties.
WS-RenewableReferences	A conventional decoration of a WS-Addressing endpoint reference with policy information needed to retrieve an updated version of an endpoint reference when it becomes invalid.
WS-ServiceGroup	An interface to heterogeneous by-reference collections of Web services.
WS-BaseFaults	A base fault XML type for use when returning faults in a Web service message exchange.

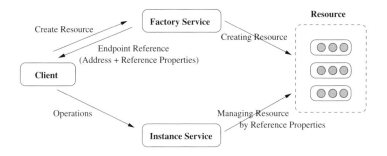

FIGURE 2.9: The implied resource pattern.

WSRF defines a family of five composable specifications (see Table 2.3) that together with the WS-Notification (see Table 2.4), which addresses event notification subscription and delivery and the WS-Addressing specifications [47] provide similar functionality to that of OGSI.

The fundamental conceptual difference between WSRF and OGSI resides in the way of modeling resources using Web services. OGSI treats a resource as a Web service itself (i.e., by supporting the GridService PortType). WSRF, on the other hand, makes explicit distinction between the "service" and the "resources" acted upon by that service by using the *implied resource pattern* [50] to describe views on state and to support its management through associated properties.

2.4.2.1 The implied resource pattern

The implied resource pattern for stateful resources refers to the mechanisms used to describe the relationship between Web services and stateful

Table 2.4: WS-Notification Specifications summary.

Specification Name	Description
WS-BaseNotification	Defines Web services operations to define the roles of notification producers and notification consumers.
WS-BrokeredNotification	Defines Web services operations for a notification broker. A notification broker is an intermediary which, among other things, allows publication of messages from entities that are not themselves service providers. It includes standard message exchanges to be implemented by notification broker service providers along with operational requirements expected of service providers and requestors that participate in brokered notifications.
WS-Topics	Defines a mechanism to organize and categorize topics. It defines three topic expression dialects that can be used as subscription expressions in *subscribe request* messages and other parts of the WS-Notification system. It further specifics an XML model for describing meta data associated with topics.

resources through a set of conventions on existing Web services technologies, particularly XML, WSDL, and WS-Addressing [47]. The term *implied* is used because the identity of the stateful resource is not specified explicitly in the request message, but rather is treated as implicit input for the execution of the message request using the reference properties feature of WS-Addressing. The endpoint reference (EPR) provides the means to point to both the Web service and the stateful resource in one convenient XML element. This means that the requestor does not provide the stateful resource identifier as an explicit parameter in the body of the request message. Instead, the stateful resource is implicitly associated with the execution of the message exchange [55].

In the implied resource pattern, a stateful resource is modeled in terms of *WS-Resource* and is uniquely identified through the EPR as illustrated in Figure 2.9. The Factory Service is capable of creating new instance services and is responsible for creating the resource, assigning it an identity, and creating a WS-Resource qualified endpoint reference to point to it. An Instance Service is required to access and manipulate the information contained in the resources associated with this service. The EPR contains, in addition to the endpoint address of the Web services, other metadata associated with the Web services such as service description information and *reference properties*, which help to define a contextual use of the endpoint reference. The reference properties of the endpoint reference play an important role in the implied resource pattern.

The framework defines how to declare, create, access, monitor for change, and destroy the WS-Resource through conventional Web services mechanisms. It describes how to make the properties of a WS-Resource accessible through a Web service interface and to manage a WS-Resources lifetime.

In the following section, we present the key management features of WSRF.

2.4.2.2 Resource representation: WS-Resource

The core WSRF specification is WS-Resource [55], which is defined as the composition of a resource and a Web service through which clients can access the state of this resource and manage its lifetime. The WS-Resource is not very restrictive with respect to what can be considered a resource. A resource has to satisfy at least two requirements: it needs to be uniquely identifiable and it must have properties. A WS-Resource uses a network-wide pointer EPR with WS-Addressing reference properties and Resource Properties to meet these requirements. The EPR with a set of WS-Addressing reference properties refers to the unique identity of the resource and the URL of the managing Web services. Resource Properties reflect the state data of the stateful resources. It should be noted that these Resource Properties could vary from simple to complex data types and even reference other WS-Resources. Referencing other Resources through Resource Properties is a powerful concept, which defines and elaborates interdependency of the WS-Resources at a lower level. A set of Resource Properties are aggregated into a resource property document: an XML document that can be available to the service requestors so that they can query it using XPath or any other query languages.

The lifetime of resource instances can be renewed before expiration as specified by the WS-ResourceLifetime specification. They can also be destroyed prematurely as required by the application. The lifetime of an instance of a resource is managed by the client itself or any other process interacting as a client, independent of the Web service and its container. It is possible for multiple Web services to manage and monitor the same WS-Resource instance with different business logic and from a different perspective. Similarly, WS-Resources are not confined to a single organization and multiple organizations may work together with the same managing Web services.

2.4.2.3 Service addressing: WS-Addressing

WSRF uses the WS-Addressing specification [47] endpoint reference construct for addressing of a WS-Resource. An endpoint reference is used to represent the address of a Web service deployed at a given network endpoint. The fact that an endpoint reference may also contain metadata associated with the Web services makes it appropriate to be used in the implied resource pattern in which a stateful resource is treated as an implied input for the processing of a message sent to a Web service. The endpoint reference construct is used to uniquely identify the stateful resource to be used in the execution of all message exchanges performed by this Web service. A WS-Resource endpoint reference may be returned as a result of operations such as a Web service message request to a factory service which instantiates and returns a reference to a new WS-Resource, from the evaluation of a search query on a service registry or as a result of some application-specific Web services request.

2.4.2.4 Resource lifetime management: WS-ResourceLifetime

The WS-ResourceLifetime specification [58] proposes mechanisms for negotiating and controlling the lifetime of WS-Resource. The specification defines a set of standard message exchange patterns for destroying, establishing, and renewing a resource, either immediately or by using a time-based scheduled mechanism. The specification also supports extension of the scheduled termination time of a WS-Resource at runtime. This feature offers explicit destruction capabilities to a service requestor. Once the resource is destroyed, the resource EPR is no longer valid and the service requestor will not be able to connect to the resource using the same EPR.

A set of service properties and two types of service destruction message patterns are defined. The service properties that are used to manage the lifetime of service are: InitialTerminationTime, CurrentTime, and TerminationTime. The identified service destruction message exchange patterns for lifetime management capabilities are immediate and scheduled destruction.

- *Immediate destruction*: allows the service requestor to explicitly request the immediate termination of a resource instance by sending an appropriate request (DestroyRequest message) to the Web services, together with the WS-Resource qualified endpoint reference. The Web service managing the WS-Resource takes the endpoint reference and identifies the specific resource to be destroyed. Upon the destruction of the resource the Web service sends a reply to the requestor with a message that acknowledges the completion of the request. Any further message exchanges with this WS-Resource will return a fault message.

- *Scheduled destruction*: allows the service requestor to define a specified period of time in the future by sending an appropriate request (SetTerminationTimeRequest message) to the Web service, together with the WS-Resource qualified endpoint reference. Using this endpoint reference, the service requestor may first establish and subsequently renew the scheduled termination time of the WS-Resource. When that time expires, the WS-Resource may be self-destroyed without the need for a synchronous destroy request from the service requestor. The requestor may periodically update the scheduled termination time to adjust the lifetime of the WS-Resource.

In addition to the above capabilities, the specification supports the notification to interested parties when the resource is destroyed through notification topics. As defined in the WS-Notification specification, the Web services associated with the WS-Resource could be a notification producer, which proposes the notification topics to allow service requestors to subscribe to notification about the destruction of a specific resource.

2.4.2.5   Resource properties: WS-ResourceProperties

The definition of the properties of a WS-Resource is standardized in the WS-ResourceProperties specification [59] as a part of the Web services interface in terms of a resource properties document. The WS-Resources properties represent a view of the resource's state in XML format.

The WS-ResourceProperties standardizes the set of message exchanges for the retrieval, modification, update and deletion of the contents of resource properties and supporting subscription for notification when the value of a resource property changes. The set of properties defined in the resource properties document associated with the service interface defines the constraints on the valid contents of these message exchanges.

2.4.2.6   Service collection: WS-ServiceGroup

It is possible to represent and manage heterogeneous collections of Web services, in order to provide a domain-specific solution, or a simple collection of services, for indexing and other discovery scenarios. The WS-ServiceGroup specification [61] defines the mechanisms for organizing a "by-reference" collection of Web services, and provides key manageability interfaces to better manage entries in the group (e.g., add, delete, and modify). Although any Web services can become a part of this collection, the service group can be used to form a wide variety of collections of Web services or WS-Resources, for example to build registries, or to build services that can perform collective operations on a set of WS-Resources.

The resource property model from WS-ResourceProperties is used to express membership rules, membership constraints, and classifications. Details of each member in the service collection are expressed through WS-ResourceProperties, which wraps the EndpointReference and the contents of the member. WS-ServiceGroup also defines interfaces for managing the membership of a ServiceGroup.

2.4.2.7   Fault management: WS-BaseFaults

Fault management is a difficult issue in Web services applications since each application uses a different convention for representing common information in fault messages. In this context, the WS-BaseFaults specification [80] defines a base fault type, which is used to return faults in a Web service message exchange. Web services fault messages declared in a common way improve support for problem identification and fault management. It enforces also the development of common tooling to assist in the handling of faults described uniformly. WS-BaseFaults defines an XML Schema type for a base fault, along with rules for how this fault type is used and extended by Web services. It standardizes the way in which errors are reported by defining a standard base fault type and procedure for use of this fault type inside WSDL. WS-BaseFault defines different standard elements corresponding to the time when the fault occurred (Timestamp), the endpoint of the Web service that generated the fault (OriginatorReference), error code (ErrorCode), error description (Description), the cause for the fault (FaultCause) and any arbitrary information required to rectify the fault.

2.4.2.8 Notification: WS-Notification

WSRF exploits the family of WS-Notification specifications, including WS-BaseNotification [62], WS-BrokeredNotification [63] and WS-Topics [64], which define a standard approach to notification using a topic-based publish and subscribe pattern. More specifically, the goal of WS-BaseNotification is to standardize exchanges and interfaces for producers and consumers of notifications. WS-BrokeredNotification aims to facilitate the deployment of Message Oriented Middleware (MOM) to enable brokered notifications between producers and consumers of the notifications. WS-Topics deals with the organization of subscriptions and defines dialects associated with subscription expressions, which are used in conjunction with exchanges that take place in WS-BaseNotification and WS-Brokered Notification.

WS-Notification currently also makes use of two other specifications in WSRF context: WS-ResourceProperties to describe data associated with resources, and WS-ResourceLifetime to manage lifetimes associated with subscriptions and publisher registrations (in WS-BrokeredNotifications).

2.4.3 OSGI vs. WSRF

As discussed previously, OGSI and WSRF are two approaches developed to enable the management of stateful resources through Web services interfaces. However, there exist some fundamental differences between these two approaches that will be described in this section. Table 2.7 outlines the mappings from OGSI concepts and constructs to equivalent WSRF concepts and constructs.

2.4.3.1 Resource modeling

OGSI differs from WSRF in the modeling of resources. While OGSI treats a stateful resource as a Web service (i.e., a grid service), WSRF makes clearer the distinction between the Web service interface and the underlying stateful resource they manage. In other words, OGSI encapsulates the state information in the grid service interface. WSRF, in the other hand, defines a separate interface containing the state information and the operations to modify it.

In OGSI, a grid service interface must declare a PortType whose type is, or is extended from GridService PortType. OGSI declares Service Data elements as part of an interface definition, which provides a standard way for querying

Table 2.5: OGSI to WS-Resource Framework and WS-Notification map [54].

OGSI	WS-Resource Framework
Grid Service Reference	WS-Addressing Endpoint Reference.
Grid Service Handle	WS-Addressing Endpoint Reference and WS-RenewableReferences.
HandleResolver PortType	WS-RenewableReferences.
Service Data Definition	Resource properties definition.
GridService PortType service data access	WS-Resource Properties.
GridService PortType lifetime management	WS-ResourceLifetime.
Notification PortTypes	WS-Notification.
Factory PortType	Now treated as a WS-Resource Factory concept.
ServiceGroup PortTypes	WS-ServiceGroup.
Base fault type	WS-BaseFault.
GWSDL	Copy-and-paste. Uses existing WSDL 1.1 interface composition approaches (that is, copy and paste) rather than using WSDL 2.0 constructs.

and accessing the state data. However, the WSDL specification version 1.1 does not allow a PortType to be extended and it is not possible to have additional information to a PortType. OGSI proposes a GWSDL PortType to overcome this limit. The fact that OGSI extends WSDL makes OGSI not compatible with existing Web services tools.

In WSRF, the term "grid services" was deprecated. Therefore, it is inappropriate to consider grid services an extension of basic Web services. The key idea separates the grid service concept of OGSI into "normal" Web services and the stateful resources that the Web services manage. The state of resources is specified through Resource Property elements, which are conceptually identical to Service Data elements and defined by standard XML Schema elements. These Resource Property elements are collected in a Resource Properties document, which is then associated with the interface of the service by using an XML attribute on the WSDL 1.1 PortType. This way of definition of services and its associated resources makes WSRF compatible with WSDL 1.1.

2.4.3.2 State information

OGSI differs from WSRF in the way the data associated with stateful resources is presented to clients. As Service Data elements define the state of the resources, OGSI proposes a set of functions for retrieving and manipulating these elements. For example, within the GridService interface, `findServiceData` is defined for returning the Service Data upon client queries, and `setServiceData` is defined for modifying or deleting a certain Service Data element.

Table 2.6: A WS-Resource-qualified Endpoint Reference.

```
<EndpointReference>
   <Address>http://host/wsrf/Service</Address>
   <ReferenceProperties>
      <ResourceKey>8807d620</ResourceKey>
   </ReferenceProperties>
</EndpointReference>
```

In WSRF, the state of resources is reflected by Resource Property elements, which are defined in a Resource Properties document. Resource Property elements can be retrieved and modified through a set of specific operations defined in the WS-ResourceProperties interface, such as GetResourceProperty, GetMultipleResourceProperties, SetResourceProperties, QueryResourceProperties.

2.4.3.3 State addressing

Since resources are created dynamically and their state may change during their lifetime, a mechanism to access the state of resources across a Web service infrastructure in an inter-operable and reliable way is needed. OGSI and WSRF take different approaches to address the state of the resources.

OGSI defines Grid Service Handle (GSH) and Grid Service Reference (GSR) as the standardized representation of grid service address. A GSH is a persistent handle assigned to the service instance, but it does not contain sufficient addressing information for a client to connect to the service instance. The GSR plays the role of a transient network pointer with associated metadata related to the grid services, such as service description information and reference properties associated with a contextual use of the targeted grid services, which can be used to locate and invoke the grid services. The GSH can be resolved to a GSR using a "Handle Resolver" mechanism. The Handle Resolver PortType defines a standard operation findByHandle, which returns one or more GSRs corresponding to a GSH. A service instance that implements the Handle Resolver PortType is referred to as a handle resolver.

In contrast, WSRF uses WS-Addressing to provide Web services endpoints and contextual identifiers for stateful resources known as WS-Resources. The Endpoint References construct defined in the WS-addressing specification is adopted as an XML structure for identifying Web services endpoints. These EndpointReferences may be returned by the factory that creates a new WS-Resource and contains other metadata such as *reference properties*. These reference properties encapsulate the stateful resource identifier that allows identifying a specific WS-Resource associated with the service. Table 2.6 shows a WS-Addressing endpoint reference as used within the conventions

Table 2.7: Mapping from OGSI to WSRF lifetime management constructs [54].

Function	OGSI	WSRF
Create new entity	Factory PortType operation "createService"	Factory pattern definition
Address the entity	Grid Service Handle and Grid Service Reference	WS-Addressing Endpoint Reference with reference properties
Immediate destruction	GridService PortType operation "destroy"	ResourceLifetime PortType operation "Destroy". However, this operation is synchronous in WSRF
Scheduled destruction	GridService PortType operations, "requestTerminationAfter" and "requestTermination-Before"	ResourceLifetime PortType operation "SetTerminationTime" is equivalent to "After". "Before" was determined to be superfluous in the absence of real-time scheduling
Determine current time	GridService PortType service data element "CurrentTime"	Resource property "CurrentTime"
Determine lifetime	GridService PortType service data element "TerminationTime"	Resource property "Termination-Time"
Notify of destruction	Not available	Subscribe to topic "ResourceDestruction"

of WS-Resource. The endpoint reference contains two components: (i) the *Address* component encapsulates the network transport-specific address of the Web service, and (ii) the *ReferenceProperties* component contains a stateful resource identifier.

The fact that WSRF exploits existing XML standards, as well emerging Web services standards such as WS-Addressing, makes it easier to implement within existing and emerging Web services toolkits, and easier to exploit within the myriad of Web services interfaces in definition.

2.4.3.4   Lifetime management

In OGSI and WSRF context, Web services are stateful and dynamic (i.e., transient); the lifetime within the services is non-trivial. Lifetime management is a crucial aspect in both OGSI and WSRF models. OGSI and WSRF manage the lifetime of their stateful resources in a slightly different way.

- *Creation*: OGSI addresses the service creation via the Factory PortType, which provides an operation "createService", that takes as optional arguments a proposed termination time and execution parameters. This operation returns a service locator for the newly created service, an initial termination time, and optional additional data.

  WSRF defines the *factory pattern*, a term used to refer to a Web service that supports an operation that creates and returns endpoint references for one or more newly created WS-Resources. In that way, the creation of a stateful Web service (i.e., grid service) in OGSI really corresponds to the creation of a WS-Resource in WSRF.

- *Destruction*: OGSI addresses destruction via operations supported in its GridService PortType, which allows the service requestor to explicitly request destruction of a grid service. OGSI proposes two operations for managing grid service lifetime including `requestTerminationAfter` and `requestTerminationBefore`.

  WSRF standardizes two approaches for the destruction of a WS-Resource: immediate and scheduled destruction. A WS-Resource can be destroyed immediately using the appropriate WS-Resource-qualified endpoint reference for the destroy request message. The service requestor may also establish and later renew a scheduled termination time of the WS-Resource. When the time expires the WS-Resource may self destruct.

2.4.3.5 Service grouping

Service grouping is a particularly important aspect when dealing with stateful entities. Both OGSI and WSRF propose a standard mechanism for creating a heterogeneous by-reference collection of services or resources. OGSI and WSRF allow grouping of service instances in essentially the same way. OGSI addresses this feature via three interfaces: the ServiceGroup interface which represents the group of grid services, the ServiceGroupEntry interface which allows management of the individual entries in a group, and the ServiceGroupRegistration interface which defines the operations to add or remove an entry to or from a group. In WSRF, the equivalent interfaces are defined in the WS-ServiceGroup specification.

The only difference between the two approaches is that the "remove" operation on the ServiceGroupRegistration interface, which allows the removal of a set of matching services, is not included in WS-ServiceGroup. This operation was removed mainly because of its redundancy with removing services from a group by doing lifetime management on the service group entry resource (i.e., the ServiceGroupEntry can be destroyed using the normal WS-ResourceLifetime operations).

2.4.3.6 Notification

In an environment in which stateful resources may change their state dynamically, it becomes important to provide support for asynchronous notification of changes.

OGSI meets this requirement via its notification interfaces, which allow a client to define a subscription (i.e., a persistent query) against one or more service data values. However, subscription and notification are broad concepts, since not all events relate to changes in the state of a service or resource.

WSRF extends the original OGSI notification model by exploiting WS-Notification. The WS-Notification family of specifications introduces a more feature-complete, generic, hierarchical topic-based approach for publish/subscribe-based notification, which is a common model followed in large scale, distributed event management systems.

2.4.3.7　Faults

WSDL defines a message exchange fault model, but not a base format for fault messages. A common base fault mechanism is a crucial requirement for common interpretation of fault messages generated by different distributed services.

OGSI addresses this issue by defining a base XML schema definition (i.e., a base XSD type, `ogsi:FaultType`) and associated semantics for fault messages, together with a convention for extending this base definition for various types of faults. By defining a common base set of information that all fault messages must contain, the identification of faults between services is simplified.

WSRF adopts the same constructs, defining them in the WS-BaseFault specification. The only difference is the removal of the open extensibility from WS-BaseFault, because it is redundant with the required approach of extending the base fault type using an XML schema extension for extended faults and because that extensibility element placed an additional burden on the capabilities of broadly available Web services tooling [49].

2.5　Concluding remarks

Initially, grid computing was defined as a hardware and software infrastructure that provides dependable, consistent, pervasive, and inexpensive access to high-end computational capabilities. The next phase of the evolution of grid systems would involve the "service paradigm" to achieve more common usage, and to provide incentives for users to wrap their existing applications as grid services. The trend toward the modeling resources as services has led to the emergence of a family of specifications, such as OGSA/OGSI, WSRF, and WS-Notification, which describes a set of services and interactions enabling implementation of a grid. These specifications enforce traditional Web services with features such as state and lifecycle, making them more suitable for managing and sharing resources on the grid.

References

[39] UDDI version 3.0.2: UDDI spec technical committee draft. Available online at: `http://uddi.org/pubs/uddi_v3.htm` (Accessed August 31st, 2007).

[40] Universal Description, Discovery and Integration (UDDI). Available online at: `http://www.uddi.org` (Accessed August 31st, 2007).

[41] The Globus alliance, Nov. 2004. Available online at: `http://www.globus.org` (Accessed August 31st, 2007).

[42] World wide web consortium (W3C): leading the web to its full potential, 2004. Available online at: `http://www.w3c.org` (Accessed August 31st, 2007).

[43] B. Allcock, J. Bester, J. Bresnahan, A. L. Chervenak, C. Kesselman, S. Meder, V. Nefedova, D. Quesnel, S. Tuecke, and I. Foster. Secure, efficient data transport and replica management for high-performance data-intensive computing. In *Proceedings of the 18th IEEE Symposium on Mass Storage Systems (MSS 2001), Large Scale Storage in the Web*, page 13, Washington, DC, USA, 2001. IEEE Computer Society.

[44] T. Banks. Web Services Resource Framework (WSRF) - Primer v1.2, May 2006. Available online at: `http://www.oasis-open.org/committees/wsrf` (Accessed August 31st, 2007).

[45] C. Barreto, V. Bullard, T. Erl, J. Evdemon, D. Jordan, K. Kand, D. Knig, S. Moser, R. Stout, R. Ten-Hove, I. Trickovic, D. van der Rijn, and A. Yiu. Web Services Business Process Execution Language Version 2.0 - Primer, May 2007.

[46] M. Bartel, J. Boyer, B. Fox, B. LaMacchia, and E. Simon. XML-Signature syntax and processing, Aug. 2001. Available online at: `http://www.w3.org/TR/2001/PR-xmldsig-core-20010820/` (Accessed August 31st, 2007).

[47] D. Box, E. Christensen, F. Curbera, D. Ferguson, J. Frey, C. Kaler, D. Langworthy, F. Leymann, B. Lovering, S. Lucco, S. Millet, N. Mukhi, M. Nottingham, D. Orchard, J. Shewchuk, E. Sindambiwe, T. Storey, S. Weerawarana, and S. Winkler. Web Services Addressing (WS-Addressing), Aug. 2004. Available online at: `http://www.w3.org/Submission/2004/SUBM-ws-Addressing-20040810/` (Accessed August 31st, 2007).

[48] E. Christensen, F. Curbera, G. Meredith, and S. Weerarawana. Web Service Description Language (WSDL). W3C note 15, Mar. 2001. Available online at: `http://www.w3.org/TR/wsdl` (Accessed August 31st, 2007).

[49] K. Czajkowski, D. Ferguson, I. Foster, J. Frey, S. Graham, T. Maguire, D. Snelling, and S. Tuecke. From Open Grid Services Infrastructure to WS-Resource Framework: refactoring & evolution, version 1.1, 2004.

[50] K. Czajkowski, D. Ferguson, I. Foster, J. Frey, S. Graham, I. Sedukhin, D. Snelling, S. Tuecke, and W. Vambenepe. The WS-Resource Framework. version 1.0, May 2004.

[51] A. Djaoui, S. Parastatidis, and A. Mani. Open grid service infrastructure primer. Technical report, Global Grid Forum, Aug. 2004. Available online at: `http://www.ggf.org/documents/GWD-I-E/GFD-I.031.pdf` (Accessed August 31st, 2007).

[52] R. Fielding, U. Irvine, J. Gettys, J. Mogul, H. Frystyk, and T. Berners-Lee. RFC-2068: Hypertext Transfer Protocol - HTTP/1.1, Jan. 1997. Available online at: `http://www.w3.org/Protocols/rfc2068/rfc2068` (Accessed August 31st, 2007).

[53] I. Foster, D. Berry, A. Djaoui, A. Grimshaw, B. Horn, H. Kishimoto, F. Maciel, A. Savva, F. Siebenlist, R. Subramaniam, J. Treadwell, and J. V. Reich. The Open Grid Services Architecture. version 1.0, July 2004.

[54] I. Foster, K. Czajkowski, D. F. Ferguson, J. Frey, S. Graham, T. Maguire, D. Snelling, and S. Tuecke. Modeling and managing state in distributed systems: the role of OGSI and WSRF. *Proceedings of the IEEE*, 93(3):604–612, 2005.

[55] I. Foster, J. Frey, S. Graham, S. Tuecke, K. Czajkowski, D. Ferguson, F. Leymann, M. Nally, I. Sedukhin, D. Snelling, T. Storey, W. Vambenepe, and S. Weerawarana. Modeling stateful resources with Web services), Mar. 2004. Available online at: `http://www.ibm.com/developerworks/library/ws-resource/ws-modelingresources.pdf` (Accessed August 31st, 2007).

[56] I. Foster and C. Kesselman. *The Grid: Blueprint for a New Computing Infrastructure.* Morgan Kaufmann, 1999.

[57] I. Foster, C. Kesselman, J. M. Nick, and S. Tuecke. Grid services for distributed system integration. *Computer*, 35(6):37–46, 2002.

[58] J. Frey, S. Graham, K. Crajkowski, D. Ferguson, I. Foster, F. Leymann, T. Maguire, N. Nagaratnam, M. Nally, T. Storey, I. Sedukhin, D. Snelling, S. Tuecke, W. Vambenepe, and S. Weerawarana.

Web Services Resource Lifetime (WS-ResourceLifetime). version 1.1, May 2004. Available online at: `http://www.ibm.com/developerworks/library/ws-resource/ws-resourcelifetime.pdf` (Accessed August 31st, 2007).

[59] S. Graham, K. Crajkowski, D. Ferguson, I. Foster, J. Frey, F. Leymann, T. Maguire, N. Nagaratnam, M. Nally, T. Storey, I. Sedukhin, D. Snelling, S. Tuecke, W. Vambenepe, and S. Weerawarana. Web Services Resource Properties (WS-ResourceProperties). version 1.1, May 2003. Available online at: `http://www.ibm.com/developerworks/library/ws-resource/ws-resourceproperties.pdf` (Accessed August 31st, 2007).

[60] S. Graham, D. Hull, and B. Muray. Web Services Base Notification 1.3 (WS-BaseNotification), Oct. 2006. Available online at: `http://www.oasis-open.org/committees/wsn` (Accessed August 31st, 2007).

[61] S. Graham, T. Maguire, J. Frey, N. Nagaratnam, I. Sedukhin, D. Snelling, K. Crajkowski, S. Tuecke, and W. Vambenepe. Web Services Resource Service Group - Specification (WS-ServiceGroup). version 1.0, Mar. 2004. Available online at: `http://www.ibm.com/developerworks/library/ws-resource/ws-servicegroup.pdf` (Accessed August 31st, 2007).

[62] S. Graham, P. Niblett, D. Chappell, A. Lewis, N. Nagaratnam, J. Parikh, S. Patil, S. Samdarshi, I. Sedukhin, D. Snelling, S. Tuecke, W. Vambenepe, and B. Weihl. Web Services Base Notification (WS-Base Notification). version 1.0, May 2004. Available online at: `ftp://www6.software.ibm.com/software/developer/library/ws-notification/WS-BaseN.pdf` (Accessed August 31st, 2007).

[63] S. Graham, P. Niblett, D. Chappell, A. Lewis, N. Nagaratnam, J. Parikh, S. Patil, S. Samdarshi, I. Sedukhin, D. Snelling, S. Tuecke, W. Vambenepe, and B. Weihl. Web Services Brokered Notification (WS-BrokeredNotification). version 1.0, May 2004. Available online at: `ftp://www6.software.ibm.com/software/developer/library/ws-notification/WS-BrokeredN.pdf` (Accessed August 31st, 2007).

[64] S. Graham, P. Niblett, D. Chappell, A. Lewis, N. Nagaratnam, J. Parikh, S. Patil, S. Samdarshi, I. Sedukhin, D. Snelling, S. Tuecke, W. Vambenepe, and B. Weihl. Web Services Topics (WS-Topics). version 1.0, May 2004. Available online at: `ftp://www6.software.ibm.com/software/developer/library/ws-notification/WS-Topics.pdf` (Accessed August 31st, 2007).

[65] A. Grimshaw and S. Tuecke. Grid services extend web services. *Web Services Journal*, 3(8):22–26, 2003.

[66] X. S. W. Group. XML Schema: Primer, 2001. Available online at: `http://www.w3.org/TR/xmlschema-0/` (Accessed August 31st, 2007).

[67] I. S. G. Heather Kreger. Web Services Conceptual Architecture (WSCA 1.0), May 2001. Available online at: `http://www.cs.uoi.gr/~zarras/ mdw-ws/WebServicesConceptualArchitectu2.pdf` (Accessed August 31st, 2007).

[68] T. Imamura, B. Dillaway, and E. Simon. Xml encryption syntax and processing, Dec. 2002. Available online at: `http://www.w3.org/TR/ xmlenc-core/` (Accessed August 31st, 2007).

[69] J. Joseph, M. Ernest, and C. Fellenstein. Evolution of grid computing architecture and grid adoption models. *IBM Systems Journal*, 43(4):624–645, 2004.

[70] A. Nadalin, C. Kaler, P. Hallam-Baker, and R. Monzillo. Web Services Security: SOAP Message Security 1.0 (WS-Security 2004), Mar. 2004. Available online at: `http://docs.oasis-open.org/wss/2004/01/ oasis-200401-wss-soap-message-security-1.0.pdf` (Accessed August 31st, 2007).

[71] E. Ort. Service-Oriented Architecture and Web services: Concepts, technologies, and tools, Apr. 2005.

[72] M. P. Papazoglou and J.-J. Dubray. A survey of web service technologies. Technical report, Department of Information and Communication Technology, University of Trento, 38050 Povo - Trento, Italy, Via Sommarive 14, June 2004.

[73] J. Pathak. Should we compare web and grid services? Available online at: `http://wscc.info/p51561/files/paper63.pdf` (Accessed August 31st, 2007).

[74] J. Postel and J. Reynolds. RFC-959: File Transfer Protocol (FTP), Oct. 1985. Available online at: `http://www.w3.org/Protocols/rfc959/` (Accessed August 31st, 2007).

[75] J. B. Postel. RFC-821: Simple Mail Transfer Protocol (SMTP), Aug. 1982. Available online at: `http://rfc.sunsite.dk/rfc/rfc821.html` (Accessed August 31st, 2007).

[76] K. Rhodes. XML-RPC vs. SOAP. Available online at: `http:// weblog.masukomi.org/writings/xml-rpc_vs_soap.htm` (Accessed August 31st, 2007).

[77] A. Skonnard. Understanding WSDL. Available online at: `http:// msdn2.microsoft.com/en-us/library/ms996486.aspx` (Accessed August 31st, 2007).

[78] L. Srinivasan and J. Treadwell. An overview of Service-Oriented Architecture, Web services and grid computing, Nov. 2005.

[79] A. TAN. Understanding the SOAP protocol and the methods of transferring binary data.

[80] S. Tuecke, K. Crajkowski, J. Frey, I. Foster, S. Graham, T. Maguire, I. Sedukhin, D. Snelling, and W. Vambenepe. Web Services Base Faults (WS-BaseFaults). version 1.0, Mar. 2004. Available online at: `http://www.ibm.com/developerworks/library/ws-resource/ws-basefaults.pdf` (Accessed August 31st, 2007).

[81] S. Tuecke, K. Czajkowski, I. Foster, J. Frey, S. Graham, and C. Kesselman. Grid service specification, 2002.

[82] S. Tuecke, K. Czajkowski, I. Foster, J. Frey, S. Graham, C. Kesselman, T. Maguire, T. Sandholm, D. Snelling, and P. Vanderbilt. Open Grid Services Infrastructure (OGSI). version 1.0, July 2003.

[83] W3C. SOAP version 1.2 part 0: Primer. W3C recommendation, June 2003. Available online at: `http://www.w3.org/TR/soap/` (Accessed August 31st, 2007).

[84] W. W. W. C. (W3C). Web services architecture. W3C working group note 11, Feb. 2004. Available online at: `http://www.w3.org/TR/ws-arch` (Accessed August 31st, 2007).

[85] W. W. W. C. (W3C). Web Service Description Language (WSDL) version 2.0 part 1: Core language. W3C working draft, May 2007. Available online at: `http://www.w3.org/TR/wsdl20/` (Accessed August 31st, 2007).

[86] D. Winer. XML-RPC Specification, June 1999. Available online at: `http://www.xmlrpc.com/spec` (Accessed August 31st, 2007).

Chapter 3

Data management in grid environments

3.1 Introduction

Grid technology enables access and sharing of computing and data resources across distributed sites. However, the grid is also a complex environment which is composed of various and heterogeneous machines. The goal of grid computing is to provide transparent access to resources in such a way that the impact on applications is minimized from internal management mechanism of the grid.

This transparency feature must be applied to access and to manage data for the execution of data-intensive applications in the grid. The emphasis lies on providing common interfaces between existing data storage systems in order to make them work seamlessly. This will not only liberate novice grid users (e.g., scientists) from data access-related issues so they may concentrate on the problems in their fields but also limit the change of interfaces between existing applications. A uniform Application Programming Interface (API) for managing and accessing data in distributed systems is needed. As a result, it is necessary to develop middleware that automates the management of data located in distributed data sources in grid environments.

3.2 The scientific challenges

In recent years, the data requirements for scientific applications have been growing dramatically in both volume and scale. Much scientific research is now data intensive. Today, information technology must cope with an ever-increasing amount of data. In the past the amount of data generated by computer simulations was usually limited by the available computational technology. The increase in archival storage was comparable to the increase in computational capability. In 1999 this view is no longer correct. What has changed is the fact that we will have to deal increasingly with experimental data, which are generated from new technologies such as high-energy physics, climate modeling, earthquake engineering, bioinformatics, and astronomy. In these domains, the volume of data for an average scientific application which was measured in terabytes has been rising to petabytes in just a couple of years. These data requirements continue to increase rapidly each year and they are expected to reach to the exabyte scale within the next decade. There are many examples that illustrate the spectacular growth of data requirements for scientific applications.

High energy physics The most cited example of massive data generation in the field of high-energy physics is the Large Hadron Collider (LHC) - the world's most powerful particle accelerator at CERN, the European Organization for Nuclear Research. Four High Energy Physics (HEP) experiments, which consist of ALICE, ATLAS, CMS and LHCb, will produce several petabytes (PB) of raw and derived data per year over a lifetime of 15 to 20 years. For example, the CMS will produce 10^9 events per seconds (1GHz). It will require fast access to approximately 1 exabyte of data, which is accumulated after the first 5 to 8 years of detector operation. These data will be accessed from different centers around the world through very heterogeneous computational resources.

The raw data are generated at a single location (CERN) where the accelerator and experiments are hosted, but the computational capacity required to analyze them implies that the analysis must be performed at geographically distributed centers. In practice, CERN's experiments are collaborations among thousands of physicists from about 300 universities and institutes in 50 countries, so the experiment's data are not only stored centrally and locally at CERN but located at worldwide distributed sites, called Regional Centers (RCs). It means that generated data need to be shared among the different user communities distributed at many sites world-wide. The computing model of a typical experiment is shown in Figure 3.1. These resources are organized into a hierarchical multi-tier grid structure. Users should have transparent and efficient access to the data, irrespective of their location. Hence, special efforts for data management and data storage are required.

These RCs are part of the distributed computing model and should complement the functionality of the CERN center. The aim is to decentralize the computing power and data storage in these RCs in order to allow physicists to do their analysis work outside of CERN with a reasonable response time rather than accessing all the data at CERN. This should also help scientists spread around the world to collaboratively work on the same data. RCs will be set up in different places around the globe.

In the HEP community, the produced data can be distinguished as raw data generated by the detector, reconstructed physics data and tag summary data. The amount of raw data produced per year will be about 1 PB. The amount of reconstructed data will be around 500 terabyte (TB) per year. The experiment

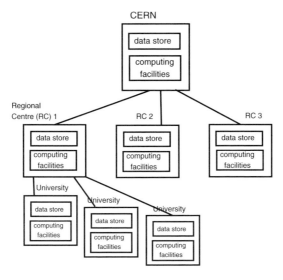

FIGURE 3.1: Example of the network of one experiment computing model.

will run for about 100 days per year, therefore roughly 5 TB of reconstructed data will be produced per day. The amount of reconstructed data to be transferred to a single RC is 200 TB per year. This mass of data will be stored and managed across multiple RCs through the LHC Computing Grid (LCG) project. Since the consumers of raw data and reconstructed physics data are distributed at many RCs worldwide, data need to be transferred efficiently between CERN and RCs. Hence, it is essential for the success of the HEP grid to have the high performance networks that should be able to transfer a massive amount of data on demand. It is also desirable to make copies or *replicas* of the data being analyzed to minimize access time and network load. Future particle accelerators like the proposed International Linear Collider (ILC) are likely to have even more intensive data requirements.

Climate modeling Another example of the growth of data requirements for scientific applications is climate model computations [120]. Climate modeling requires long-duration simulations and generates very large files that are needed to analyze the simulated climate. The goal is to take a statistical ensemble mean of the simulation results in suppressing the growth of errors included in the initial observational data and those generated during the simulation. The computations execute hundreds to thousands of sample simulations while introducing perturbation for each simulation. The result of each sample simulation is gathered and included in an average to generate the final statistical result.

As the complexity and size of the simulations grow, the volume of generated model output threatens to outpace the storage capacity of current archival systems and transfering it across distributed sites faces challenges. As an example, a high-resolution computational ocean model running on computers with peak speeds in the 100-gigaflop range can generate a dozen multi-gigabyte files in a few hours at an average rate of about 2 MB/second. Computing a century of simulated time takes more than a month to complete and produces about 10 TB of archival output. Archival systems capable of storing hundreds of terabytes are required to support calculations of this scale. Moving to one-teraflop systems and beyond requires petabyte archival systems.

Earth observation An example in the field of earth science, the Earth Observing System (EOS) is a program of NASA including a series of artificial satellite missions and scientific instruments in Earth orbit designed for long-term global observations of the land surface, biosphere, atmosphere, and oceans of the Earth. Sensed data about the Earth captured by various NASA and non-NASA satellites are transferred to various archives. Then, the archives extract calibrated and validated geophysical parameters from the raw data. For this purpose, NASA developed eight Distributed Active Archive Centers (DAACs), which are intended to hold and distribute long-term Earth observation data from the EOS. The DAACs are a significant component of the Earth Observing System Data and Information System (EOSDIS). Since raw data in the EOSDIS DAACs contain little scientific interest, they are carefully transformed into calibrated and validated data.

Calibration, validation and production of customized data require significant resources. The DAACs currently have about 3 to 4 PB of data and provide data to more than 100,000 customers per year. They distribute many TB per week spread around the world according to user orders through a website (more than two million distinct IP addresses accessed the web interfaces of the EOSDIS data centers in 2004). Currently, the ordered data products from DAACs are delivered via an FTP (file transfer protocol) site or via media. FTP is chosen for transfer if the order is small (less than a few gigabytes). Tape or CD-ROM delivered via the postal service is used for larger data orders.

Bioinformatic Genomics require programs such as genome sequencing projects, which are producing huge amounts of data. The analysis of these raw biological data requires very large computing resources. Bioinformatics involves the integration of computers, software tools, and databases in an effort to address these biological applications. Genome sequences provide copious information about species from microorganisms to human beings. The analysis and comparison of genome sequences are necessary for the investigation of genome structures which is useful for the predictions about the functions and activities of organisms.

As an example, in applications such as design of new drugs, large databases are required for extensive comparison and analyses of genome sequences. As the rate of complete genome sequencing is continually increasing, genome comparison and analysis have become data intensive tasks. There is a growing need for capacity storage and effective transfer of genome data.

Astronomy Another example of data intensive applications in the astronomy field is the Sloan Digital Sky Survey (SDSS) [106], which aims to map in detail one quarter of the entire sky and determines positions and absolute brightness of more than 100 million celestial objects. It will also measure the distances to more than a million galaxies.

Astronomical applications are performed in several regions of the electromagnetic spectrum and produce an enormous amount of data. Usually, the map of a particular region of the sky is obtained by several groups using different techniques to generate a two-dimensional image. These images are manipulated so that they can be compared and overlapped. Furthermore, it is necessary to compare images obtained in different wave-lengths. All this manipulation is made pixel by pixel and requires a considerable computational power to construct this atlas of the firmament and to implement an online database with the collected material. The data volume produced nowadays is about 500TB per year in images that should be stored and made available for all the researchers in the field. Starting in 2008, the Large Synoptic Survey Telescope should produce more than 10PB per year.

3.3 Major data grid efforts today

3.3.1 Data grid

The "data grid" has been considered a unifying concept to describe the new technologies offering a comprehensive solution to data intensive applications. Data grid services encapsulate the underlying network storage systems and provide a uniform interface to applications. With these services, users can discover, transfer, and manipulate shared datasets stored in distributed repositories and also create and manage copies of these datasets. At least, a data grid provides two basic functionalities: a high performance, reliable data transfer mechanism, and a scalable replica discovery and management mechanism [203]. Depending on application requirements, various other services need to be provided, such as consistency management for replicas, metadata management. All operations in a data grid are mediated by a security layer that handles authentication of entities and ensures conduct of only authorized operations. Data grids are typically characterized by:

- *Large-scale*: They consist of many resources and users across distributed sites.

- *Service oriented environment*: They propose new services on top of existing local mechanisms and interfaces in order to facilitate the coordinated sharing of remote resources.

- *Uniform and transparent access*: They provide user applications with transparent access to computing and data resources: computer platforms, file systems, databases, collections, data types and formats, as well as computational services. This transparency is vitally important for sharing heterogeneous distributed resources in a manageable way.

- *Single point of authentication/authorization*: They provide users with a single point of authentication/authorization to access data holdings from distributed sites, based on user access rights, and authorize shared access to data holdings across sites, while maintaining strict levels of privacy and security, auditing mechanisms may be also available.

- *Consistency of replicas*: They can seamlessly create data replicas and maintain their consistency, to ensure quality of service, including fault tolerance, disaster recovery and load balancing.

The world of grid computing is continuously growing, more concretely many new data grid projects are founded in an increasing rate. These projects have been initiated by the near-term needs of scientific experiments in various different fields and have led to collaborations between scientific communities and computer communities. This collaboration allows scientists from various disciplines partnering with computer scientists to develop and exploit production-scale data grids. Table 3.1 contains a list of some of the major data grid projects, which are described in this section. This list is not exhaustive. The projects have been chosen based on several attributes that are relevant to the development of the projects, such as domains, application environment and tools, project status, etc.

3.3.2   American data grid projects

3.3.2.1   GriPhyN

The Grid Physics Network (GriPhyN) project [99] funded by the National Science Foundation (NSF) develops one of the most advanced concepts for data management in various physics experiments. While the short term goal is to combine the data from the CMS and ATLAS experiments at the LHC (Large Hadron Collider), LIGO (Laser Interferometer Gravitational Observatory) and SDSS (Sloan Digital Sky Survey), the long-term goal of this project is to deploy petabyte-scale computational environments based on technologies and experience from developing GriPhyN to meet the data intensive needs of a community of thousands of scientists spread across the globe.

Table 3.1: List of data grid projects summarized in this section.

Name	Domains	Country	Remarks
GriPhyN [NSF, 2000-2005]	High energy physics	United States	To deploy PB-scale computational environments to meet the data-intensive needs of a community of thousands of scientists spread across the globe.
PPDG [DOE, 1999]	High energy physics	United States	Having close collaboration with the GriPhyN and the CERN DataGrid with the long-term goal of forming a Petascale Virtual-Data Grid.
iVDGL [NSF, 2001-2006]	High energy physics	United States	To construct computational and storage infrastructure based on heterogeneous computing and storage resources from the US, Europe, Asia, Australia and South America via high-speed networks.
TeraGrid [NSF, 2001]	High energy physics, biology	United States	To create the grid infrastructure that interconnects some of the US's fastest supercomputers.
DataGrid [European Union, 2001-2004]	High energy physics, earth science and biology	Europe	To create a grid infrastructure to provide online access to data on a petabyte scale.
DataTAG [European Commission, 2002-2004]	Network	Europe and the United States	To provide a global high performance intercontinental grid testbed based on a high speed transatlantic link connecting existing high-speed GRID testbeds in Europe and the US.
CrossGrid [European Union, 2002]	High-energy physics, biomedical and earth science	Europe	To extend a grid environment across European countries and to new application areas.
GridPP [PPARC, 2002]	High energy physics	United Kingdom	To create computational and storage infrastructure for particle physics in the UK.
Network GÉANT [European Commission, 2000]	Network	Europe	To develop the GÉANT network - a multi-gigabit pan-European data communications network, reserved specifically for research and education use.
EGEE [European Union, 2004-2005]	High energy physics, biomedical sciences	Europe	To create a seamless common grid infrastructure to support scientific research.
LHC Computing Grid project [Industry and CERN, 2005]	High energy physics	Europe	To create and maintain data movement and analysis infrastructure for the users of LHC.

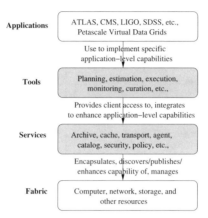

FIGURE 3.2: Architecture of the Virtual Data Toolkit.

Clients and domains   GriPhyN is a collaboration of both experimental physicists and information technology researchers.

Application environment and tools   GriPhyN focuses on realizing the concepts of Virtual Data, which involve in developing new methods to catalog, characterize, validate, and archive software components to implement virtual data manipulations. Moreover, Virtual Data aims to provide a virtual space of data products derived from experimental data, in which requests for data products are transparently mapped into computation and/or data access operations across multiple grid computing and data locations. To address this challenge, GriPhyN implements a grid software distribution *Virtual Data Toolkit (VDT)*, which consists of a set of virtual data services and tools to support a wide range of virtual data grid applications.

Figure 3.2 presents the multi-tier architecture of VDT. Applications and data grids are built based on virtual data tools, which rely on a variety of virtual data services. These services encapsulate the low-level details of hardware fabric used by the data grid. The virtual tools and virtual services can integrate components developed in existing grid middlewares (e.g., Condor, MCAT/SRB, Globus toolkit, etc.) to fulfill a specific functionality such as parallel I/O, high-speed data movement, authentication and authorization.

3.3.2.2   Particle Physics Data Grid (PPDG)

The Particle Physics Data Grid (PPDG) created in 1999 [104] is a collaboration between computer scientists and physicists among universities to develop, evaluate and deliver distributed data access and management infrastructures for large particle and nuclear physics experiments.

Clients and domains The PPDG project [104] takes a major role in the international coordination of grid projects relevant to high-energy and nuclear physics fields. Especially, it has very close collaboration with the GriPhyN and the CERN DataGrid with the long-term goal of combining these efforts to form a Petascale Virtual-Data Grid. PPDG proposes novel mechanisms and policies including the vertical integration of grid middleware with experiment-specific applications and computing resources to form effective end-to-end capabilities.

Application environment and tools The PPDG has adopted the Virtual Data Toolkit (VDT), which was initially developed by GriPhyN and supported by iVDGL-like software packaging and distribution mechanism of the common grid middleware. The VDT is also used as the underlying middleware for the European physics-focused grid projects including the Enabling Grids for EsciencE (EGEE) and Worldwide LHC Computing Grid.

The PPDG aims to provide a data transfer solution with additional functionalities for file replication, file caching, pre-staging, and status checking of file transfer requests. These capabilities are constructed on the existing functionalities of the SRB, Globus, and the US LBNL HPSS Resource Manager (HRM). Security is provided through existing grid technologies such as GridFTP and Grid Security Infrastructure (used by SRB).

Project status The project is currently ongoing with funds approved by SciDAC. In 2005, PPDG joined with the NSF-funded iVDGL, US LHC Computing Grid project, DOE Laboratory facility and other groups to build, operate and extend their systems and applications on the production Open Science Grid.

3.3.2.3 International Virtual Data Grid Laboratory (iVDGL)

The International Virtual Data Grid Laboratory (iVDGL), which was formed in 2001 is constructed on heterogeneous computing and storage resources from the U.S., Europe, Asia, Australia and South America via high-speed networks. The iVDGL enables the international collaborations for interdisciplinary experimentation in grid-enabled data intensive scientific computing. More concretely, laboratory users will be able to realize scientific experiments from various projects such as gravitational wave searches projects (e.g., Laser Interferometer Gravitational-wave Observatory - LIGO), high energy physics projects (e.g., the ATLAS and CMS detectors at the Large Hardon Collider - LHC at CERN), digital astronomy projects (e.g., the Sloan Digital Sky Survey - SDSS), and the U.S. National Virtual Observatory (NVO).

3.3.2.4 TeraGrid

The TeraGrid [108] is a large project launched by NSF in August 2001. TeraGrid refers to the infrastructure that interconnects some of the US's fastest supercomputers with high-speed storage systems and visualization equipment at geographically dispersed locations.

Clients and domains   The primary goal of the TeraGrid project is to provide a grid infrastructure with an unprecedented increase in the computational capabilities both in terms of capacity and functionality dedicated to open scientific research. TeraGrid aims to deploy a distributed "system" using grid technologies allowing users to map applications across the computational, storage, visualization, and other resources as an integrated environment.

TeraGrid envisions the following projects to use their grid computing resources:

- The MIMD Lattice Computation (MILC) collaboration

- NAMD - simulation of large biomolecular systems

Application environment and tools   TeraGrid utilizes the middleware of the NSF Middleware Initiative (NMI), which is based on the Globus toolkit. TeraGrid has support for MPI, BLAS and VTK.

Job submission and scheduling   TeraGrid uses GT's GRAM for job submission and scheduling.

Security   GSI is used for authentication.

Resource management   TeraGrid utilizes Condor for job queuing, scheduling, and for resource monitoring.

Data management   TeraGrid employs SRM for storage allocation, Globus' Global Access to Secondary Storage (GASS) for simplification of data access and GridFTP for data transfer.

Fabric   The project is constructed through a combination of several stages within the NSF TeraScale initiative. In 2000, NSF funded the TeraScale Computing System (TCS-1) at the Pittsburgh Supercomputer Center, resulting in a six teraflop computational resource. Then, in 2001, NSF funded the Distributed Terascale Facility (DTF), which is in the process of creating a fifteen teraflop computational grid composed of major resources at Argonne National Laboratory (ANL, managed by the University of Chicago), the California Institute of Technology (Caltech), the National Center for Supercomputing Applications (NCSA), and the San Diego Supercomputer Center (SDSC). The DTF grid deployed exclusively Intel's Itanium processor-based clusters distributed across the four sites. In 2002, NSF initiated the Extensible Terascale

Facility (ETF) that combines TSC-1 and DTF resources into a single 21-teraflops grid and supports heterogeneity among computational resources.

Beginning initially with four large-scale, Itanium-based Linux clusters at ANL, Caltech, NCSA, and SDSC, the TeraGrid achieved its first full-scale deployment in 2004. There are currently eight sites providing services to the network: SDSC, NCSA, ANL, PSC, Indiana University, Perdue University, Oakridge National Laboratory, and the Texas Advanced Computing Center (TACC). Among these sites, there are 16 computational systems providing more than 42 teraflops of computing power, and online storage systems offering over a petabyte of disk space via a wide area implementation of IBM's General Parallel File System (GPFS). There are also 12 PB of archival storage and a number of databases, such as the Nexrad Precipitation database at TACC, as well as science instruments and visualization facilities, such as the Quadrics Linux cluster at PSC.

3.3.3 European data grid projects

3.3.3.1 European Data Grid

The European Data Grid (EDG) project [95] funded by the European Commission was started in 2001 to join several national initiatives across the continent and in US.

Clients and domains The principal objectives of the project are to develop the software to provide basic grid functionality and associated management tools for a large scale testbed for demonstration projects in three specific areas of science including high-energy physics (HEP), earth observation and biology.

The DataGrid project focuses initially on the needs for capability of simulation and analysis of a large volume of data for each of the Large Hadron Collider experiments (ATLAS, CMS and LHCb). Recently, Earth observation science (e.g., satellite images) and the biosciences, principally genome data access and analysis, began receiving attention somewhat after HEP.

The DataGrid project is led by CERN together with five other main partners and fifteen associated partners. Apart from CERN, the main partners in the HEP part of the project are Italy's Istituto Nazionale di Fisica Nucleare (INFN), France's Centre National de la Recherche Scientifique (CNRS), UK's Particle Physics and Astronomy Research Council (PPARC), and the Dutch National Institute for Nuclear Physics and High Energy Physics (NIKHEF). The European Space Agency has taken the lead in the Earth Observation task and KNMI (Netherlands) is leading the biology and medicine tasks. In addition to the major partners, there are associated partners from the Czech Republic, Finland, Germany, Hungary, Spain and Sweden. A relatively recent important development is the establishment of formal collaboration with some of the US grid projects (e.g., GriPhyN and PPDG projects).

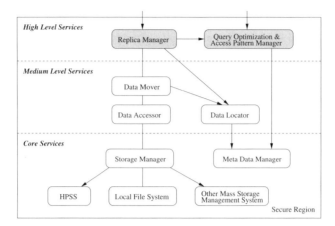

FIGURE 3.3: European DataGrid Data Management architecture.

Application environment and tools   Figure 3.3 shows the Data Management architecture proposed for the European DataGrid. The *Replica Manager* manages files and meta data copies in a distributed and hierarchical cache with a specific replication policy. It further uses the *Data Mover* to accomplish its tasks. The *Data Mover* takes care of transferring files from one storage system to another one. To implement its functionality, it uses the *Data Accessor* and the *Data Locator*, which map location-independent identifiers to location-dependent identifiers. The *Data Accessor* is an interface encapsulating the details of the local file system and mass storage systems such as Castor, HPSS and others. The *Data Locator* makes use of the generic *Meta Data Manager*, which is responsible for efficient publishing and management of a distributed and hierarchical set of associations between meta data and its data. *Query Optimization* and *Access Pattern Management* ensure that for a given query an optimal migration and replication execution plan is produced. Such plans are generated on the basis of published meta data including dynamic logging information. All components provide appropriate security mechanisms that transparently span worldwide independent organizational institutions. The granularity of access is both on the file level as well as on the data set level. A data set is seen as a set of logically related files.

Fabric   The work is divided into twelve work packages: Grid Workload Management (WP1), Grid Data Management (WP2), Grid Monitoring Services (WP3), Fabric Management (WP4), Mass Storage Management (WP5), Integration Testbed (WP6), Network Services (WP7), HEP Applications (WP8), Earth Observation Science Applications (WP9), Biology Applications (WP10), Dissemination (WP11), Project Management (WP12).

The first five of these packages will each develop specific well-defined parts of the grid middleware. The Testbed & Network (WP6, WP7) activities will integrate the middleware into a production quality infrastructure linking several major laboratories spread across Europe, providing a large scale testbed for scientific applications. The others are related to applications in earth science, satellite remote sensing and biology.

3.3.3.2 DataTAG

The DataTAG project [92] complements EDG by providing a global high performance intercontinental grid testbed based on a high speed transatlantic link connecting existing high-speed GRID testbeds in Europe and USA. The DataTAG established new records in long-distance data transfers via international networks. Then, this project was superseded by the Enabling Grids for E-science project (Section 3.3.3.4), which has constructed production-quality infrastructure and built the largest multi-science grid in the world, with over 200 sites.

Clients and domains The DataTAG testbed focuses upon advanced networking issues and inter-operability between the intercontinental grid domains, hence extending the capabilities of each and enhancing the worldwide program of grid development.

Application environment and tools The DataTAG project has many innovative components in the area of high performance transport, Quality of Service (QoS), advance bandwidth reservation, EU-US Grid inter-operability and new tools for easing the management of Virtual Organizations such as the Virtual Organization Membership Server (VOMS) and grid monitoring (GridICE). Together with DataGrid and the LHC Computing Grid (LCG) project, the software of the CERN LHC experiments ALICE, ATLAS CMS and LHCb has been adapted to the grid environment.

3.3.3.3 European Research Network GÉANT

The GÉANT project [97] launched in November 2000 was a collaboration between 26 National Research and Education Networks (NRENs) representing 30 countries across Europe, the European Commission, and DANTE. DANTE is the project's coordinating partner.

Clients and domains The project's principal purpose was to develop the GÉANT network - a multi-gigabit pan-European data communications network, reserved specifically for research and education use. This network is based on the previous TEN-155 pan-European research network. The project also covered a number of other activities relating to research networking. These included network testing, development of new technologies and support for some research projects with specific networking requirements.

Application environment and tools In addition to the development of the GÉANT network, the project also covers a number of other activities relating to research networking. These include network testing, development of new technologies and support for other related projects.

Fabric Currently, GÉANT network has 12Gbps connectivity to North America, and 2.5Gbps to Japan. Additional connections to GÉANT have been established to the Southern Mediterranean through the EUMEDCON-NECT project. Work is also underway to establish additional connections to GÉANT for NRENs from other world regions, including Latin America (through ALICE) and the Asia-Pacific region (through TEIN2).

3.3.3.4 Enabling Grids for E-science in Europe

The Enabling Grids for E-science in Europe (EGEE) project [94] launched in April 2004 is a European Grid project that aims to provide computing resources to European academia and industries. Working areas include the implementation of a European grid infrastructure, development and maintenance of grid middleware and training and support of grid users. Many of its activities are based on experiences from the EDG project (Section 3.3.3.1).

Clients and domains EGEE aims to provide researchers in both academia and industries with access to major computing resources, independent of their geographic location. The main applications for EGEE are the LHC experiments. EGEE has chosen as pilot projects LCG and Biomedical Grids.

Application environment and tools The middleware integrates middleware from the VDT, the EDG and the AliEN project.

Fabric The infrastructure of the EGEE computation grid is built on the EU Research Network GÉANT and national research and education networks across Europe. The amount of CPUs has grown from 3000 CPUs at the beginning of the project to over 8000 by the end of the second year.

3.3.3.5 LHC Computing Grid project

The LHC Computing Grid (LCG) project [101] is building and maintaining a grid infrastructure for the high energy physics community in Europe, USA and Asia. The main purpose of the LCG is to handle the massive amounts of data produced from the LHC (Large Hadron Collider) experiments at CERN.

Clients and domains The main applications for LCG are high energy physics, biotechnology and other applications that EGEE brings in. The main application is the gathering of data from the LHC experiments ATLAS, CMS, Alice and LHCb.

Fabric The amount of computers the centers participating in the LCG project have to manage is so massive that manual maintenance of the installed software of these computers is too labor intensive. Also, with such a number of components, the failure of one component should automatically be overridden and not affect the overall operability of the system. The LCG project has designed fabric management software, which automates some of these tasks.

3.3.3.6 CrossGrid

The CrossGrid project [88] formed in 2002 aims to extend a grid environment across European countries and to new application areas.

Clients and domains The CrossGrid project aims to develop, implement and exploit new grid components for interactive compute-intensive and data-intensive applications, including simulation and visualization of surgical procedures, flooding simulations, team decision support systems, distributed data analysis in high-energy physics, air pollution and weather forecasting. The project, with partners from eleven European countries, will also install grid testbeds in a user-friendly environment to evaluate and validate the elaborated methodology, generic application architecture, programming environment and new grid services. The Cross Grid is closely working with the Grid Forum and the EU DataGrid project to profit from their results and experience, and to obtain full inter-operability. This collaboration intends to extend the grid across eleven European countries.

Application environment and tools The CrossGrid project plans to build a software grid toolkit, which will include tools for scheduling and monitoring resources.

3.3.3.7 GridPP

The GridPP project is developing a computing grid for particle physics, in a collaboration with particle physicists and computer scientists from the UK and CERN.

Clients and domains GridPP grid is intended for applications in particle physics. More concretely, it focuses on creating a prototype grid involving four main areas: (i) support for the CERN LHC Computing Grid (LCG), (ii) middleware development as part of the European DataGrid (EDG), (iii) the development of particle physics applications for the LHC and US experiments, and (iv) the construction of grid infrastructure in the UK.

Application environment and tools GridPP contributes to middleware development in a number of areas, mainly through the EGEE project [119].

An interface to the APEL accounting system (Accounting Processor for Event Logs: an implementation of grid accounting which parses log files to extract and then publish job information) has also been provided and is being tested. The development of the R-GMA monitoring system has continued, with improvements to the stability of the code and robustness of the system deployed on the production grid. A major re-factored release of R-GMA was made for gLite-1.5. Similarly, GridSite was updated where it provides containerized services for hosting VO (Virtual Organization) boxes (machines specific to individual virtual organizations that run VO-specific services such as data management: an approach which, in principle, is a security concern) and support for hybrid HTTPS/HTTP file transfers (referred to as "GridHTTP") to the htcp tool used by EGEE. GridSiteWiki has been developed, which allows Grid Certificate access to a wiki, preventing unauthorized access, and which is in regular use by GridPP. The cornerstone of establishing a grid is a well-defined security policy and its implementation: GridPP leads the development of that security policy within EGEE, having identified 63 vulnerability issues at the end of 2005. Monitoring and enhancements of the networking, workload management system (WMS) and data management systems have been performed in response to deployment requirements, with various tools developed (e.g., GridMon for network performance monitoring), Sun Grid Engine integration for the WMS, and MonAMI, a low-level monitoring daemon integrated with various data management systems.

3.4   Data management challenges in grid environments

While the challenges on the computing side are already quite tremendous, supercomputer centers must also cope with an ever-increasing amount of data with the emergence of data intensive applications. This engenders access and movement of very large data collections among geographically distributed sites. These collections consist of raw and refined data ranging in size from terabytes to petabytes or more. While standard grid infrastructures provide users with the ability to collaborate and share resources, special efforts concerning data management and data storage are needed to respond to the specific challenges raised by data intensive activities.

In this section we point out the main requirements that pose challenges for data management in grid environments.

Data namespace organization   A problem for data sharing in a heterogeneous storage system is data namespace organization. The reason is that each storage system has its own mechanism for naming the resources. Resource naming affects other resource management functions such as resource discovery. The data management system needs to define a logical namespace in which every data element has a unique logical filename. The logical filename is mapped to one or more physical filenames on various storage resources across distributed storage systems in the grids.

Transparent access to heterogeneous data repositories One of the fundamental problems that any data management system needs to address is the heterogeneity of repositories where data are stored. This aspect becomes even more challenging when data management has to be targeted in grid environments, which spread over multivirtual organizations in a wide area network environment. The main reason is the variety of possible storage systems, which can be multiple disk storage systems like DPSS, distributed file systems like AFS, NFS, or even databases. This diversity imposes the way in which the data sets are named and accessed. For example, data are identified through a file name in distributed file systems, or through an object identifier in databases.

The high level applications should not need to be aware of the specific low-level mechanisms required to access data in a particular storage system. They should be presented with a uniform view of data and with uniform mechanisms for accessing that data. Hence, data management systems should provide a component service, which defines a single interface for higher level applications to access data located in different underlying repositories. The role of this component is to make the appropriate conversions for grid data access requests to be executed in the underlying storage system. This component service hides from higher layers the complexities and specific mechanisms for data access, which are particular to each storage system.

Efficient data transfer In scientific applications, data are normally stored at a central place. Scientists who would like to work with the data need to make local copies of parts of the data. The job of data management systems is to deal with large amounts of data (terabytes or petabytes) that have to be transferred over the wide area networks. Hence, there is an essential requirement for efficient data transfer between sites. At present, there are already emerging some enhanced FTP variants, such as Globus GridFTP and CERN RFIO (Remote File I/O) for data transfers in the grids.

RFIO is developed as a component of the CERN Advanced Storage Manager (Castor), which implements remote versions of most POSIX calls like *open*, *read*, *write*, *lseek* and *close*, and several Unix commands like *cp*, *rm*, and *mkdir*. RFIO provides libraries to access files on remote disks or in the Castor namespace. GridFTP is a high-performance, secure protocol using GSI (Grid Security Infrastructure) for authentication, and having useful features such as TCP buffer sizing and multiple parallel streams. It is enhanced with a variety of features to be used as a tool for higher-level application data access on the grid.

Data replication   Replication can be considered as the process of managing identical copies of data at different places in a grid environment. It is desirable for data to be replicated at different sites to minimize access time and network load by allowing user applications access to local cached data stores rather than to transfer each single requested file over the wide area network to the application. Replication is also needed for fault tolerance and this requirement effects the "efficient data transfer" requirement above.

One of the main issues in data replication is the consistency of replicas. Since a replica is not just a simple copy of an original but still has a logical connection to it, it is important to maintain the consistency between replicas. Data consistency depends on how frequently the data is updated and the amount of data items covered by the update, so the consistency problem is more complicated when updates are possible on replicas. However, the key problem of data replication is not only the update mechanisms in order to guarantee the consistency among the different replicas, but also related to policies or strategies that should be applied for replica creation. The reason is that in a grid environment it is impossible to impose a single replication policy for every participating site. For example, system administrators can decide for production requirements to distribute data according to some specific strategies, and job schedulers may require specific data replication policies to speed up execution of jobs. Hence, data management systems need to provide appropriate services for various types of users (e.g., grid administrators, job schedulers) to be able to replicate, maintain consistency and obtain information about replicas.

Data security   In a distributed grid environment, access to data should obviously be controlled. The security of data being transferred over wide area networks should be ensured. Allowing data to be exchanged without some form of encryption makes it possible for secure data to be read as it is transferred over public networks. Equally, data storage should be handled in such a way that ensures that the data cannot be read by unauthorized people or applications.

Moreover, encryption of stored data with public/private keys, using the security of the operating system, and using authentication to prevent malicious data from being introduced must be implemented in the data management systems for the grid environment, but should be monitored by the grid manager to ensure that the data is secure at all times. Encryption keys should be regularly updated, and solutions should be regularly tested and verified for correct data, especially in a distributed grid environment.

Another key security issue of data management in wide area networks is related to data caches. It is necessary to maintain the same level of security between the participating sites. For example, the site that owns the original data needs to ensure that the remote sites holding replicas of its data provide the same level of security as the owner requires for their data. This becomes a critical issue when it is about sensitive data where human or intellectual rights exist. The fact that each site may use different security architecture makes this task more complicated.

3.5 Overview of existing solutions

Current data management solutions for the grid environment are largely based on four approaches. In this section, we summarize the major data management solutions in each approach.

3.5.1 Data transport mechanism

Data transport in grids can be modeled as a three-tier structure [147]: *transfer protocol* as the bottom layer, *overlay network* as the second layer and *application-specific* as the top layer. The first layer specifies the transport protocols for data movement between two nodes in a network, such as FTP, GridFTP. The second layer aims to provide the routing mechanism for the data and services such as storage in the network, caching of data transfers for better reliability, and the ability for applications to manage transfer of large datasets. An overlay network provides a specific semantic over the transport protocols to satisfy a particular purpose. The topmost layer provides applications with transparent access to remote data through APIs that hide the complexity and the unreliability of the networks.

Initial efforts to manage data on the grid are based primarily on explicit data movement methods. These methods concentrate to develop file transfer protocols, which actually move data between machines in a grid environment, and overlay mechanisms for distributing data resource across Data Grids.

3.5.1.1 Transfer protocols

There exist a number of protocols such as FTP, HTTP for transferring files between different machines. However, they are not adapted for the grid. Therefore, the lack of standard protocols for transfer and access of data in the grid has led to a fragmented grid storage community. Users who wish to access different storage systems are forced to use multiple protocols and/or APIs, and it is difficult to efficiently transfer data among these different storage systems. In the context of the Globus project, a common data transfer and access protocol called GridFTP [111] that provides secure, efficient data movement in grid environments is proposed. This protocol, which extends the standard FTP (File Transfer Protocol) protocol, provides the extended features in order to support data transfers in the grid. GridFTP allows using parallel data transfer through multiple TCP streams to improve bandwidth over using a single TCP stream. It supports third-party control of transfers between storage servers, striped data transfer, partial file transfer, etc. Moreover, GridFTP is based on Grid Security Infrastructure (GSI), which provides a robust and flexible authentication, integrity, and confidentiality mechanism for transferring files. UberFTP [110] is the first interactive, GridFTP client. GSI-OpenSSH is a modified version of OpenSSH that adds support for GSI authentication and credential forwarding (delegation), providing a single sign-on remote login and file transfer service in the grid. Reliable File Transfer Service (RFT) is an OGSA-based service that provides interfaces for controlling and monitoring third party file transfers between FTP and GridFTP servers.

Apart from FTP, HTTP, and GridFTP, there exist various protocols for data transfer such as Chirp [87], Data Link Client Access Protocol (DCAP) [124], DiskRouter [132], etc. It should be noted that some middleware, such as [228] and [229], propose to use BitTorrent [125] as a protocol for large file transfer in the context of desktop grids.

3.5.1.2 Overlay mechanism

The overlay mechanism approach [114], [115] for data management focuses on optimization of data transfer and storage operations for a globally scalable, maximally inter-operable storage network environment. This storage-enabled network environment allows data to be placed not only in computer-center storage systems but also within a network fabric enhanced with temporary storage. Data transfers between two nodes can be optimized by controlling data transfer explicitly by storing the data in a temporary buffer at intermediate nodes. Applications can manipulate these buffers so that data is moved to locations close to where it is required. The key point to notice in this network is that services of various kinds can be provided to data stored in transit at the intermediate nodes. This infrastructure defines a framework with basic storage services upon which higher level services can be created to meet user needs. In this network, some scheduling models for data transfers can be considered to be applied in conjunction with scheduling models of computational jobs, such as [143].

Based on an overlay mechanism approach, the IBP project [136], [113] provides a general store-and-forward overlay networking infrastructure. IBP is modeled after the Internet Protocol. It defines a networking stack that is similar to the OSI reference model for large-scale data management in distributed networks. We present in the following section the networking stacks proposed by IBP (Fig 3.4).

Internet Backplane Protocol (IBP). IBP storage servers are machines installed with IBP server software, called *depots*. IBP depots allows clients to perform remote storage operations, such as storage management, data transfer and depot management. The lowest layer of the storage net-

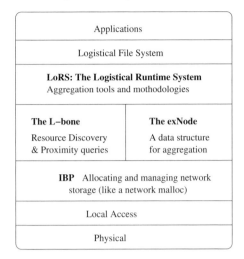

| Applications |
| Logistical File System |
| **LoRS: The Logistical Runtime System** Aggregation tools and mothodologies |

| **The L–bone** Resource Discovery & Proximity queries | **The exNode** A data structure for aggregation |

| **IBP**   Allocating and managing network storage (like a network malloc) |
| Local Access |
| Physical |

FIGURE 3.4: The Network Storage Stack.

working stack is the Internet Backplane Protocol (IBP), which defines a mechanism to share storage resources across networks ranging from LAN to WAN, and it allows applications to control the storage, data, and the data transmission between IBP depots. From the view of clients, a depot's storage resources are a collection of append-only byte arrays. A chief design feature of IBP is the use of *capability*, which is cryptographically secure byte strings generated by the IBP depot. The capabilities are assigned by depots and they can be viewed as the handles of the byte arrays. Capabilities provide a notion of security as the client has to use the same capabilities to perform the subsequent operations.

Logistical Backbone (L-Bone).   The L-Bone layer allows clients access to a collection of IBP depots deployed in the Internet. The L-Bone server maintains a directory of registered depots in the Internet. The basic L-Bone service is to discover IBP depots, where clients can query the L-Bone for depots that meet certain requirements (e.g., available storage, time limits, proximity to desired hosts, and so on), and the L-Bone returns lists of candidate depots. The L-Bone uses information such as IP address, country code, and zip code to determine proximity for the depots.

external Nodes (exNodes).   Following the example of the *inode* concept in the Unix file-system, the exNode is designed to manage aggregate allocations on network storage. In a IBP network, a large data file can be aggregated from multiple IBP byte arrays stored on different IBP servers. An exNode is the collection of capabilities of allocated IBP byte-arrays. The exNode library handles IBP capabilities and allows the user to associate metadata with the capabilities. The exNode library has a set of functions that allow an application to create and destroy an exNode, to add or delete a mapping from it, to query it with a set of criteria, and to produce an XML serialization. When a user wants to store the exNode to disk or to pass it to another user, he can use the exNode library to serialize it to an XML file. With this file, users can manage the corresponding allocated storage in IBP.

Logistical Runtime System (LoRS). Although the L-Bone makes it easier for the user to find depots and the exNode handles IBP capabilities for the user, the user still has to manually request allocations, store the data, create the exNode, attach mappings to the exNode and insert the IBP allocations and metadata into the mappings. The LoRS layer consists of a C-API and a command line interface tool set, which can automatically find IBP depots via the L-Bone, operate IBP capabilities, and create exNodes. The LoRS facilitates the operations on network files in IBP.
IBP follows an approach that relies on explicit data management, which provides no interface for transparent access to data. Besides, guarantee of data persistence and consistency is at the user's charge. The objective of IBP is to provide a low-level storage solution that functions just above the networking layer upon which higher level services can be built to provide transparent access to data. As an example, IBP has been used for data management in Grid-RPC Netsolve [112] to create an infrastructure that enables the efficient and robust servicing of distributed computational requests with large data requirements. Other projects that follow the similar approach to IBP are presented briefly in the following section.

Globus Access to Secondary Storage (GASS) [117] is provided within the Globus Toolkit and implements a variety of data access strategies, enabling programs running at various locations to read and write remote data through a uniform remote I/O interface. GASS uses special Uniform Resource Locators (URLs) to identify data stored in remote file systems on the grid. These URLs may be in the form of an HTTP URL (if the file is accessible via an HTTP server) or an x-gass URL (in other cases). From the users' point of view, using GASS does not differ very much from using files from the local file system. The only difference is that GASS provides new functions to open and close files (i.e., *gass_fopen* and *gass_fclose*) but after that GASS files behave exactly like any other file: they can be read and written using the standard file I/O operations. When an application requests a remote file for reading, GASS fetches the remote file into a cache from where it is opened for reading. The cache is maintained as long as applications are accessing it. When an application wants to write to a remote file, the file is created or opened within the cache where GASS keeps track of all the applications writing to it via reference count.

When the reference count is zero, the file is transferred to the remote machine. In that way, all operations on the remote file are conducted locally in the cache, which reduces demand on bandwidth. GASS behaves like a distributed file system but the naming mechanism, which is based on URLs, enables it to provide efficient replica and caching mechanisms. In addition, GASS takes care of secure data transfer and authentication as well.

Kangaroo [145] proposes also a storage network of identical servers, each providing temporary storage space for a data movement service. Kangaroo improves the throughput and reliability for large data transfers within the grid. Kangaroo removes the burden of data movement from the application by handling the data transfer as a background process so that failures due to server crashes and network partitions are handled transparently by the process. In that way, the transfer of data can be performed concurrently with the execution of an application. The design of Kangaroo is similar to that of IBP even though their aims are different. Both of them use a store-and-forward method as a means of transporting data. However, while IBP allows applications to explicitly control data movement through a network, Kangaroo aims to keep the data transfer hidden through the usage of background processes. Also, IBP uses byte arrays, whereas Kangaroo uses the default TCP/IP datagrams for data transmission.

NeST [116] addresses the storage resource management by providing a mechanism for ensuring allocation of storage space in a similar way to IBP. NeST provides a generic data transfer architecture that supports multiple data transfer protocols: HTTP, FTP, GridFTP, NFS, and Chirp. The original point in NeST design is that it can negotiate with data servers to choose the most appropriate protocol for any particular transfer (e.g., NFS locally and GridFTP remotely) and optimize transfer parameters (e.g., number of parallel data flows, TCP parameters).

3.5.2 Logical file system interface

Another approach for data management in grid environments is to build a logical file-system interface based on distributed underlying file systems. Typically, this approach involves constructing the data management services providing a file-system interface offering a common view of storage resources distributed over several administrative domains, which is similar to the interface of NFS [140] for distributed file system in local network. These systems emphasize the necessary mechanisms for locating a data file in response to requests of applicative processes, such as copyTo(). The goal is to allow existing applications to access data in heterogeneous file systems without any modification in their code by providing a file access interface. A variety of techniques have been used to achieve this goal, such as interception of a system call in a C library, modifying the kernel. In this section we present a case study of GFarm, which is an existing distributed file system for grid environments.

GFarm [144] is an implementation of the Grid Datafarm architecture designed to handle hundreds of terabytes to petabytes of data using a global distributed file system. Gfarm focuses on a grid file system that provides scalable I/O bandwidth and scalable parallel processing by integrating many local file systems and clusters of thousands of nodes. It uses a metadata management system to manage the file distribution, file system metadata and parallel process information. The nodes in GFarm architecture are connected via a high speed network. In GFarm, a file is stored throughout the file system as fragments on multiple nodes. Each fragment has arbitrary length and can be stored on any node. Individual fragments can be replicated, and the replicas are managed through Gfarm metadata and replica catalog. Metadata is updated at the end of each operation on a file. A GFarm file is write-once, that is, if a file is modified and saved, then internally it is versioned and a new file is created. The core idea of GFarm is to move computation to the data. Gfarm targets data-intensive applications, which consist of independent multitasks. In these applications, the same program is executed over different data files and where the primary task is reading a large body of data. The data is split up and stored as fragments on the nodes. While executing a program, the process scheduler dispatches it to the node that has the segment of data that the program wants to access. If the nodes that contain the data and its replicas are under heavy CPU load, then the file system creates a replica of the requested fragment on another node and assigns the process to it. In this way, I/O bandwidth is gained by exploiting the access locality of data. Gfarm targets applications such as high-energy physics where the data is write-once read-many. For applications where the data is constantly updated, there could be problems with managing the consistency of the replicas and the metadata though an upcoming version aims to fix them.

GridNFS [130] is a similar middleware solution as GFarm that extends distributed file system technology and flexible identity management techniques to meet the needs of grid-based virtual organizations. The foundation for data sharing in GridNFS is NFS version 4 [141], the IETF standard for distributed file systems that is designed for security, extensibility, and high performance. NFSv4 offers new functionalities such as enhanced security, migration support, etc. The primary goal of GridNFS is to provide transparent data access in a secure way based on a global namespace offered by NFS.

LegionFS [150] is designed as file system infrastructure for grid environments. Its design is based on Legion, an object-based, user-level infrastructure for local-area and wide-area heterogeneous computation. File resources organized as Legion objects are copied into Legion space in order to support global data access.

Google File System (GFS) [128] is a scalable storage solution as it has been successfully implemented in a very large cluster. GFS is designed to provide fixed block size support for concurrent operations, and focuses on providing support for large blocks of data being read and written continuously on a distributed network of commodity hardware.

FreeLoader framework [146] The overall architecture of Freeloader shares many similarities with GFS. Freeloader aims to aggregate unused desktop storage space and I/O bandwidth into a shared cache/scratch space, for hosting large, immutable datasets and exploiting data access locality. It is designed for large data sets, e.g., outputs of scientific simulation results.

SRBfs SRBfs is based on the FUSE project [96] to provide a user-space file system interface. FUSE allows redirecting system calls of standard kernel-level file systems to a user-level library. The advantage of this technique is that developing new file systems with FUSE is relatively simple without any modification at the kernel level, providing transparency to applications. FUSE is integrated in Linux version 2.6.14. As a result, FUSE is not dedicated to file systems in grid environment. However, it allows building file systems, which provides applications transparent access to data in the grid.

3.5.3 Data replication and storage

Attempting to move large volumes of scientific data leads to a highly loaded network. When data are moved over wide-area networks, the difficulty is not only in having sufficient bandwidth but also in dealing with transient errors in the networks and the source and destination storage systems. A technique for avoiding repetitive data movement is replication of selected subsets of the data in multiple sites. Therefore, replica management is an important issue that needs to be addressed for data management in grid environments. GT provides a suite of services for replica management: MetaData Catalog Service (MCS), Replica Location Service (RLS), and Data Replication Service (DRS) [122]. These services are implemented using the Lightweight Directory Access Protocol (LDAP) [129] or databases such as MySQL.

MetaData Catalog Service (MCS)   is an OGSA-based service that provides a mechanism for storing and accessing descriptive metadata and allows users to query for data items based on desired attributes. Metadata, which is information that describes data, exists in various types. Some metadata relate to the physical characteristics of data objects, such as their size, access permissions, owners and modification information. Replication metadata information describes the relationship between logical data identifiers and one or more physical instances of the data. Other metadata attributes describe the contents of data items, allowing the data to be interpreted.

Replica Location Service (RLS)   Giggle [123] is an architectural framework for a RLS that exclusively contains metadata information related to data replication by keeping track of one or more copies, or replicas, of files in the grid environment. Data location on physical storage systems can be found through its logical name. The main goal of RLS is to reduce access latencies for applications obtaining data from remote sources and to improve the data availability thanks to their replications.

Data Replication Service (DRS) [122]   is constructed based on lower-level grid data services, including RFT and RLS services. The main function of DRS is to replicate a specified set of files onto a local storage system and register the new files in appropriate catalogs. The operations of the DRS include discovery, identifying where desired data files exist on the grid by querying the RLS. Then, the desired data files are transferred to the local storage system efficiently using the RLS service. Finally, data location mappings are registered to the RLS so that other sites may discover newly-created replicas. Throughout DRS replication operations, the service maintains state about each file, including which operations on the file have succeeded or failed.

These catalog-based services can be used to build other higher level data management services depending on user needs. For example, the Grid Data Management Pilot (GDMP) project [139], which is a collaboration between the EDG [95], [131] (in particular the Data Management work package [90]) and PPDG [104], has developed its services for data management based on Globus's catalog-based services. The project proposes a generic file replication tool that replicates files securely and efficiently from one site to another in a data grid environment. In addition, it manages replica catalog entries for file replicas and thus maintains a consistent view on names and locations of replicated files. The GDMP package has been used in the EU data grid project as well as in some high energy physics experiments in Europe and the U.S. The successor of GDMP is Reptor [134] which defines services for management of data copies. The most recent development in the EU data grid has been the edg-replica-manager [93] which makes partial use of the Globus replica management libraries for file replication. The edg-replica-manager can be regarded as a prototype for Reptor. Lightweight Data Replicator (LDR) [102] is a data management system built on top of Globus's standard data services such as GridFTP, RLS and MCS.

Another example of a high level data management system is Don Quijote [118], which is developed as a proxy service that provides management of data replicas across three heterogeneous grid environments used by ATLAS scientists: the US Grid3, the NorduGrid and the LCG-2 Grid. Each grid uses different middleware, including different underlying replica catalogs. Don Quijote provides capabilities for replica discovery, replica creation and registration, and replica renaming after data validation. Other examples of scientific grid projects that have developed customized, high-level data management services based on replica catalogs are Optor [135] and GridLab [98].

Many initiatives to build another type of high level data management system, in which the replica management services are tightly coupled with the underlying storage architecture to provide uniform access to different storage systems, such as relational databases, XML databases, file systems, etc. were undertaken by different groups of reseachers from different institutions. SRB and OGSA-DAI are typical examples of such systems.

Storage Resource Broker (SRB) [105] is a data management system for grids using a client-server architecture including three components: the Metadata Catalog (MCAT) service, SRB servers and SRB clients. SRB provides a uniform and transparent interface to access data stored in heterogeneous data storage over a network including mass storage system (e.g., High Performance Storage System [100], Data Migration Facility [91]), file systems (e.g., Unix FS, Windows FS) and databases (e.g., Oracle, DB2, Sysbase). The SRB provides an application program interface (API) which enables applications to access data stored at any of the distributed storage sites. The SRB API provides the capability to discover information, identify data collections, and select and retrieve data items that may be distributed across a Wide Area Network (WAN).

The combination of all SRB servers, clients and storage systems is called a *federation*. Every federation must have a central master server connected to a Metadata Catalog (MCAT). SRB uses MCAT service to store metadata information for the stored datasets, which allows access to data sets and resources based on their attributes rather than their names or physical locations. The SRB server consists of one or more SRB Master daemon processes with SRB Agent processes that are associated with each Master. The clients authenticate to the SRB Master, which starts an Agent process that processes the client requests. An SRB agent interfaces with the MCAT and the storage resources to execute a particular request. The fact that client requests are handed over to the appropriate server depending on the location determined by the MCAT service improves both availability of data and access performance. SRB organizes data items as collections implemented using logical storage resources (LSRs) to ensure transparency for data access and transfer. LSRs own and manage all of the information required to describe the data independent of the underlying storage system.

SRB is one of the most widely used data management systems in various data grid projects around the world, such as UK eScience Data Grid, NASA Information Power Grid, and NPACI Grid Portal Project [137].

OGSA-DAI (Data Access and Integration) [109] is the implementation of the DAIS (Data Access and Integration Services) specification [89] proposed by GGF working group. It focuses on specifying an interface, which provides location transparency to data distributed in heterogeneous storage systems in grids including relational databases, XML databases, and file systems.

3.5.4 Data allocation and scheduling

The last approach for data management in grid environments relies on the creation of systems that focus on data allocation and scheduling jobs in order to minimize the movement of data and hence the total execution time of jobs. Some typical works in this direction are listed in the following.

Stork [133] is a data placement scheduler which aims to make data placement activities first class citizens in the grid just like the computational jobs. Stork allows data placement jobs to be scheduled, monitored, managed, and even check-pointed while providing multiple transfer mechanisms (e.g., FTP, GridFTP, HTTP, DiskRouter, and NeST) and retries in the event of transient failures. As Stork is now integrated into the Condor system, Stork jobs can be managed with Condor's workflow management software (DAGMan).

Grid Application Development Software (GrADS) [126] introduces a three-phase scheduling strategy, which involves an initial matching of application requirements and available resources (launch-time scheduling), making modifications to that initial matching to take into account dynamic changes in the system availability or application changes (rescheduling), and finally coordinating all schedules for multiple applications running on the same grid at once (metascheduling).

Decoupled scheduling architecture [138] is proposed for data-intensive applications and considers data allocation and job scheduling together. The system consists of three components: the External Scheduler (ES), the Local Scheduler (LS), and the Dataset Scheduler (DS). The ES is modeled to distribute jobs to specific remote computing sites, the LS is used to decide the priority of the jobs arriving at the local node, and the DS dynamically creates replicas for popular data files. Various combinations of scheduling and replication strategies are evaluated with simulations. The simulation results show that the data locality is an important factor when scheduling the jobs. The best performance is achieved when the jobs are assigned to the sites containing the required data files and the popular datasets are replicated dynamically. Otherwise, the worst performance is given by same job scheduling strategy but without data replication. This is predictable since a few sites which host the data were overloaded in this case.

3.6   Concluding remarks

The complexity of scientific computing problems leads to an explosive demand for grid computing. This chapter first presents the challenges in terms of support for data-intensive applications as the volume and scale of data requirements for these applications increase. As a result, grid technology has evolved to meet these data requirements, which is vital for projects on the frontiers of science and engineering, such as high energy physics, climate modeling, earth observation, bioinformatics, and astronomy. We present the main grid activities today in data-intensive computing including major data grid projects on a worldwide scale. In order to effectively provide solutions for data management in grid environments, various issues need to be considered, such as data namespace organization, a mechanism for transparent access to data resources, and efficient data transfer. Finally, an overview of existing solutions for managing data in grid environments is provided.

References

[87] Chirp protocol specification. Available online at: http://www.cs.wisc. edu/condor/chirp/PROTOCOL (Accessed August 31st, 2007).

[88] The CrossGrid project website. Available online at: http://www. eu-crossgrid.org/ (Accessed August 31st, 2007).

[89] DAIS working group. Available online at: http://forge.gridforum.org/ projects/dais-wg (Accessed August 31st, 2007).

[90] Data management work package in EDG website. Available online at: http://edg-wp2.web.cern.ch/edg-wp2/ (Accessed August 31st, 2007).

[91] Data Migration Facility (DMF). Available online at: http://www.sgi. com/products/storage/tech/dmf.html (Accessed August 31st, 2007).

[92] The DataTAG project website. Available online at: http://datatag. web.cern.ch/datatag/ (Accessed August 31st, 2007).

[93] edg-replica-manager 1.0. Available online at: http://www.gridpp.ac.uk/ wiki/EDG_Replica_Manager (Accessed August 31st, 2007).

[94] The EGEE project website. Available online at: http://public. eu-egee.org/ (Accessed August 31st, 2007).

[95] European DataGrid project website. Available online at: http://www. eu-datagrid.org (Accessed August 31st, 2007).

[96] Filesystem in Userspace (FUSE). Available online at: http://fuse. sourceforge.net (Accessed August 31st, 2007).

[97] The GEANT project website. Available online at: http://www.geant. net/ (Accessed August 31st, 2007).

[98] GridLab: A grid application toolkit and testbed. Available online at: http://www.gridlab.org (Accessed August 31st, 2007).

[99] GriPhyN - grid physics network website. Available online at: http: //www.griphyn.org/ (Accessed August 31st, 2007).

[100] High Performance System Storage (HPSS). Available online at: http: //www.hpss-collaboration.org (Accessed August 31st, 2007).

[101] The LCG project website. Available online at: http://lcg.web.cern. ch/LCG/ (Accessed August 31st, 2007).

[102] Lightweight data replicator. Available online at: http://www.lsc-group. phys.uwm.edu/LDR/ (Accessed August 31st, 2007).

[103] MySQL website. Available online at: `http://www.mysql.com` (Accessed August 31st, 2007).

[104] Particle Physics Data Grid collaboration website. Available online at: `http://www.ppdg.net/` (Accessed August 31st, 2007).

[105] SDSC Storage Resource Broker website. Available online at: `http://www.npaci.edu/DICE/SRB/` (Accessed August 31st, 2007).

[106] Sloan digital sky survey website. Available online at: `http://www.sdss.org/` (Accessed August 31st, 2007).

[107] Storage Resource Management working group. Available online at: `http://sdm.lbl.gov/srm-wg` (Accessed August 31st, 2007).

[108] TeraGrid website. Available online at: `http://www.teragrid.org/` (Accessed August 31st, 2007).

[109] The OGSA-DAI Project website. Available online at: `http://www.ogsadai.org.uk` (Accessed August 31st, 2007).

[110] UberFTP website. Available online at: `http://dims.ncsa.uiuc.edu/set/uberftp/` (Accessed August 31st, 2007).

[111] B. Allcock, J. Bester, J. Bresnahan, A. L. Chervenak, C. Kesselman, S. Meder, V. Nefedova, D. Quesnel, S. Tuecke, and I. Foster. Secure, efficient data transport and replica management for high-performance data-intensive computing. In *Proceedings of the 18th IEEE Symposium on Mass Storage Systems (MSS 2001), Large Scale Storage in the Web*, page 13, Washington, DC, USA, 2001. IEEE Computer Society.

[112] D. C. Arnold, S. S. Vah, and J. Dongarra. On the convergence of computational and data grids. *Parallel Processing Letters*, 11(2–3):187–202, June 2001.

[113] A. Bassi, M. Beck, G. Fagg, T. Moore, J. Plank, M. Swany, and R. Wolski. The Internet Backplane Protocol: A study in resource sharing. In *Cluster Computing and the Grid 2nd IEEE/ACM International Symposium CCGRID2002*, pages 180–187, 2002.

[114] M. Beck, Y. Ding, T. Moore, and J. S. Plank. Transnet architecture and logistical networking for distributed storage. In *Workshop on Scalable File Systems and Storage Technologies (SFSST)*, San Francisco, CA, USA, Sept. 2004. Held in conjunction with the 17th International Conference on Parallel and Distributed Computing Systems (PDCS-2004).

[115] M. Beck and T. Moore. Logistical networking: a global storage network. *Journal of Physics: Conference Series*, 16(1):531–535, 2005.

[116] J. Bent, V. Venkataramani, N. LeRoy, A. Roy, J. Stanley, A. Arpaci-Dusseau, R. Arpaci-Dusseau, and M. Livny. Flexibility, manageability, and performance in a grid storage appliance. In *Proceedings of the 11th IEEE Symposium on High Performance Distributed Computing (HPDC 11)*, pages 3–12, Edinburgh, Scotland, UK, July 2002. IEEE Computer Society.

[117] J. Bester, I. Foster, C. Kesselman, J. Tedesco, and S. Tuecke. GASS: A data movement and access service for wide area computing systems. In *Proceedings of the 6th workshop on I/O in parallel and distributed systems (IOPADS '99)*, pages 77–88, Atlanta, GA, USA, 1999. ACM Press.

[118] M. Branco. Don Quijote - data management for the ATLAS automatic production system. In *Proceedings of Computing in High Energy and Nuclear Physics (CHEP)*, Interlaken, Switzerland, Sept. 2004.

[119] D. Britton, A. Cass, P. Clarke, J. Coles, A. Doyle, N. Geddes, J. Gordon, R. Jones, D. Kelsey, S. Lloyd, R. Middleton, D. Newbold, and S. Pearce. Performance of the UK Grid for particle physics. In *Proceedings of IEEE06 Conference*, Amsterdam, Dec. 2006. IEEE Computer Society. on behalf of the GridPP collaboration.

[120] A. Chervenak, E. Deelman, C. Kesselman, B. Allcock, I. Foster, V. Nefedova, J. Lee, A. Sim, A. Shoshani, B. Drach, D. Williams, and D. Middleton. High-performance remote access to climate simulation data: A challenge problem for data grid technologies. *Parallel Computing*, 29(10):1335–1356, 2003.

[121] A. Chervenak, I. Foster, C. Kesselman, C. Salisbury, and S. Tuecke. The data grid: Towards an architecture for the distributed management and analysis of large scientific datasets. *Journal of Network and Computer Applications*, 23(3):187–200, 2000.

[122] A. Chervenak, R. Schuler, C. Kesselman, S. Koranda, and B. Moe. Wide area data replication for scientific collaborations. In *GRID '05: Proceedings of the 6th IEEE/ACM International Workshop on Grid Computing*, pages 1–8, Washington, DC, USA, 2005. IEEE Computer Society.

[123] A. L. Chervenak, E. Deelman, I. T. Foster, L. Guy, W. Hoschek, A. Iamnitchi, C. Kesselman, P. Z. Kunszt, M. Ripeanu, R. Schwartzkopf, H. Stockinger, K. Stockinger, and BrianTierney. Giggle: a framework for constructing scalable replica location services. In *Proceedings of the 2002 ACM/IEEE conference on Supercomputing*, pages 1–17, Baltimore, Maryland, USA, Nov. 2002.

[124] S. T. Chiang, J. S. Lee, and H. Yasuda. Data link switching client access protocol. IETF Request For Comment 2114, NetworkWorking Group.

[125] B. Cohen. Incentives build robustness in BitTorrent. In *Proceedings of the 1st Workshop on Economics of Peer-to-Peer Systems*, Berkeley, CA, USA, June 2003.

[126] H. Dail, H. Casanova, and F. Berman. A decoupled scheduling approach for the grads environment. In *Proceedings of the IEEE/ACM SC2002 Conference (SC'02)*, Baltimore, Maryland, November 2002. IEEE.

[127] C. Ernemann and R. Yahyapour. *Grid Resource Management - State of the Art and Future Trends*, chapter Applying Economic Scheduling Methods to Grid Environments, pages 491–506. Kluwer Academic Publishers, 2003.

[128] S. Ghemawat, H. Gobioff, and S.-T. Leung. The Google file system. *SIGOPS Operating Systems Review*, 37(5):29–43, 2003.

[129] J. Hodges and R. Morgan. Lightweight Directory Access protocol (v3): Technical specification. IETF Request For Comment 3377, Network-Working Group.

[130] P. Honeyman, W. A. Adamson, and S. McKee. GridNFS: global storage for global collaborations. In *Proceedings of the IEEE International Symposium Global Data Interoperability - Challenges and Technologies*, Sardinia, Italy, June 2005. IEEE Computer Society.

[131] W. Hoschek, J. Jean-Martinez, A. Samar, H. Stockinger, and K. Stockinger. Data management in an international data grid project. In *Proceedings of the 1st IEEE/ACM International Workshop on Grid Computing (Grid '00)*, volume 1971, pages 77–90, Bangalore, India, Dec. 2000. Springer.

[132] G. Kola and M. Livny. Diskrouter: A flexible infrastructure for high performance large scale data transfers. Technical report cs-tr-2003-1484, University of Wisconsin-Madison Computer Science Department, Madison, WI, USA, 2003.

[133] T. Kosar and M. Livny. Stork: Making data placement a first class citizen in the Grid. In *ICDCS '04: Proceedings of the 24th International Conference on Distributed Computing Systems (ICDCS'04)*, pages 342–349, Washington, DC, USA, 2004. IEEE Computer Society.

[134] P. Kunszt, E. Laure, H. Stockinger, and K. Stockinger. Advanced replica management with Reptor. In *Proceedings of the 5th International Conference on Parallel Processing and Applied Mathematics*, Czestochowa, Poland, Sept. 2003.

[135] P. Kunszt, E. Laure, H. Stockinger, and K. Stockinger. File-based replica management. *Future Generation Computing Systems*, 21(1):115–123, 2005.

[136] J. Plank, M. Beck, W. Elwasif, T. Moore, M. Swany, and R. Wolski. The Internet Backplane Protocol: Storage in the network. In *Network Storage Symposium (NetStore '99)*, pages 59–59, Seattle, USA, Oct. 1999. ACM Press.

[137] A. Rajasekar, M. Wan, R. Moore, W. Schroeder, G. Kremenek, A. Jagatheesan, C. Cowart, B. Zhu, S.-Y. Chen, and R. Olschanowsky. Storage resource broker - managing distributed data in a grid. *Computer Society of India Journal, Special Issue on SAN*, 33(4):42–54, Oct. 2003.

[138] K. Ranganathan and I. T. Foster. Simulation studies of computation and data scheduling algorithms for data grids. *Journal of Grid Computing*, 1(1):53–62, 2003.

[139] A. Samar and H. Stockinger. Grid Data Management Pilot (GDMP): A tool for wide area replication in high-energy physics. In *Proceedings of the 19th IASTED International Conference on Applied Informatics (AI '01)*, Innsbruck, Austria, Feb. 2001.

[140] R. Sandberg, D. Goldberg, S. Kleiman, DanWalsh, and B. Lyon. Design and implementation of the Sun Network file system. In *Proceedings of the USENIX Summer Technical Conference*, pages 119–130, Portland, OR, USA, June 1985.

[141] S. Shepler, B. Callaghan, D. Robinson, R. Thurlow, C. Beame, M. Eisler, and D. Noveck. Network File System (NFS) version 4 protocol, 2003. RFC 3530.

[142] A. Shoshani, A. Sim, and J. Gu. Storage resource managers: Middleware components for grid storage. In *Proceedings of the 10th NASA Goddard Conference on Mass Storage Systems and Technologies, 19th IEEE Symposium on Mass Storage Systems (MSST '02)*, pages 209–223, College Park, MA, USA, Apr. 2002. IEEE Computer Society.

[143] M. Swany. Improving throughput for grid applications with network logistics. In *SC '04: Proceedings of the 2004 ACM/IEEE conference on Supercomputing*, page 23, Washington, DC, USA, 2004. IEEE Computer Society.

[144] O. Tatebe, N. Soda, Y. Morita, S. Matsuoka, and S. Sekiguchi. Gfarm v2: A grid file system that supports high-performance distributed and parallel data computing. In *Proceedings of the 2004 Computing in High Energy and Nuclear Physics (CHEP04)*, Interlaken, Switzerland, Sept. 2004.

[145] D. Thain, J. Basney, S.-C. Son, and M. Livny. The Kangaroo approach to data movement on the grid. In *Proceedings of the 10th IEEE International Symposium on High Performance Distributed Computing (HPDC*

10), pages 325–333, Francisco, CA, USA, Aug. 2001. IEEE Computer Society.

[146] S. S. Vazhkudai, X. Ma, V. W. Freeh, J. W. Strickland, N. Tammineedi, and S. L. Scott. FreeLoader: Scavenging desktop storage resources for scientific data. In *SC '05: Proceedings of the 2005 ACM/IEEE conference on Supercomputing*, page 56, Washington, DC, USA, 2005. IEEE Computer Society.

[147] S. Venugopal, R. Buyya, and K. Ramamohanarao. A taxonomy of data grids for distributed data sharing, management, and processing. *ACM Computing Surveys*, 38(1):3, 2006.

[148] B. Wei, G. Fedak, and F. Cappello. Collaborative data distribution with BitTorrent for computational desktop Grids. In *ISPDC '05: Proceedings of the 4th International Symposium on Parallel and Distributed Computing (ISPDC'05)*, pages 250–257, Washington, DC, USA, 2005. IEEE Computer Society.

[149] B. Wei, G. Fedak, and F. Cappello. Scheduling independent tasks sharing large data distributed with BitTorrent. In *GRID '05: Proceedings of the 6th IEEE/ACM International Workshop on Grid Computing*, pages 219–226. IEEE Computer Society, 2005.

[150] B. S. White, M. Walker, M. Humphrey, and A. S. Grimshaw. LegionFS: a secure and scalable file system supporting cross-domain high-performance applications. In *Proceedings of the 2001 ACM/IEEE Conference on Supercomputing (SC '01)*, pages 59–59, New York, USA, Nov. 2001. ACM Press.

[151] C.-E. Wulz. CMS - concept and physics potential. In *Proceedings II-SILAFAE*, San Juan, Puerto Rico, 1998.

Chapter 4

Peer-to-peer data management

4.1 Introduction

In recent years, peer-to-peer (hereafter P2P) networks and systems have attracted increasing attention from both the academy and industry. P2P systems are distributed systems that operate without centralized global control in the form of a global registry, global services, global resource management, global schema or data repository. In the P2P model, all participant nodes (i.e, *peers*) have identical responsibilities and are organized into an *overlay network* which is a virtual topology created on top of - and independently from - the underlying physical (typically IP) network. Each peer takes both the role of client and server. As a client, it can consume resources offered from other peers, and as a server it can provide its services for others.

Most of the time, there exists confusion between P2P systems and grid computing [211]. Although there is significant similarity between these two systems, they have some fundamental differences on their working environments. Grid systems are composed of powerful dedicated computers and CPU farms that coordinate in a large-scale network with high bandwidth based on persistent, standards-based service infrastructures. Unlike grid systems, P2P systems consist of regular user computers with slow network connections. Therefore, P2P systems suffer more failures than grid systems. As a result, P2P systems focus on dealing with instability, volatile populations, fault tolerance, and self-adaptation. Moreover, getting P2P applications to inter-operate is impossible due to a lack of common protocols and standardized infrastructure. In [161], authors claim "Grid computing addresses infrastructure, but not yet failure, while P2P addresses failure, but not yet infrastructure".

Most popular P2P systems are file or content sharing applications such as Napster, Gnutella, Chord, and CAN. They can be classified into two types of overlays: unstructured and structured. Most of the unstructured overlay networks have two characteristics. First, the distribution of resources is not based on any kind of knowledge of the network topology. There is no precise control over the network topology and the resource's location. Second, queries to find a resource are flooded across the overlay network with limited scope. Upon receiving a query, each peer queries its neighbors, which themselves query their own neighbors and so on, for a specific number of steps. Hence, an unstructured system seems to not scale well as resource location requires an exhaustive search over the network. In contrast to unstructured systems, the primary focus of structured overlay networks is on precisely locating a given resource. In other words, resources are placed at specified locations and not at random nodes. This tightly controlled structure enables the system to satisfy queries in an efficient manner.

4.2 Defining peer-to-peer

4.2.1 History

The Internet was originally conceived in the late 1960s as a P2P system. The goal of the original ARPANET was to share computing resources around the US. The challenge for this effort was to integrate different kinds of distributed resources, existing in different networks, within one common network architecture that would allow every host to be equal in terms of the functionality and tasks they perform. The first few hosts on the ARPANET including several US universities (e.g., UCLA, SRI and USCB, and the University of Utah) were already independent computing sites with equal status. The ARPANET connected them together not in a master-slave or client-server relationship but rather as equal computing peers.

As the Internet exploded in the mid 1990s, more and more computers that lack resources and bandwidth, such as desktop PCs, became clients of this network. They could not be resource providers of the system. In this context, the client-server model became more prevalent because servers provided the effective means of supporting large numbers of clients with limited resources. Most early distributed applications, such as FTP or Telnet, can be considered client-server.

In the network environment dominated by PC clients, the first wide use of P2P seems to have been in instant messaging systems such as AOL Instant Messenger and Yahoo Messenger. At the end of 1998, a 19-year-old student at Boston University, Shawn Fanning, wrote a program that allowed exchange of audio files in mp3 format across the Internet. Fanning, whose pseudonym is Napster, assigned this name to his application [160]. The originality of Napster is based on the fact that the file sharing is decentralized. The actual transfer of files is done directly between the peers. The introduction of Napster has driven the current phase of interest and activity in P2P. In fact, this peer-to-peer principle followed earlier approaches of the Internet whose goal was to create a symmetric system for sharing information.

4.2.2 Terminology

Throughout the P2P literature, there are a considerable number of different definitions of P2P systems. In fact, P2P systems are often determined more by the external perception of the end-user than their internal architecture. As a result, different definitions of P2P systems are proposed to accommodate different types of such systems. According to a widely accepted definition for P2P in the late 1990s, "P2P is a class of applications that takes advantage of resources - storage, cycles, content, human presence - available at the edges of the Internet" [190]. Munindar P. Singh attempts to describe P2P systems more extensively, rather than in just an application-specific way, and defines P2P simply as the opposite of client-server architectures [191]. The Intel P2P working group defines P2P as "the sharing of computer resources and services by direct exchange between systems" [152]. According to HP laboratory, "P2P is about sharing: giving to and obtaining from the peer community. A peer gives some resources and obtains other resources in return" [179].

In its purest form, P2P is a totally distributed system in which all nodes are completely equivalent in terms of functionality and tasks they perform. This definition is not generalized enough to embrace systems that employ the notion of "supernodes" (e.g., Kazaa) or systems that completely rely upon centralized servers for some functional operations (e.g., instant messaging systems, Napster). These systems are, however, widely accepted as P2P systems. Therefore, we propose the following definition [189]:

DEFINITION 4.1 *P2P systems are distributed systems where its participants share a part of their own hardware resources such as processing power, storage capacity, content, network link capacity. Such systems are capable of self-adapting to failures and transient status of the participants without the intermediation or support of a global centralized server or authority.*

4.2.3 Characteristics

There exist a great number of P2P systems where goals may be incompatible. However, some common characteristics that a system should possess in order for it to be termed as P2P systems are:

- *Scalability*: In client-server architecture, it is difficult to improve the scalability of the system with a considerably small cost. An immediate benefit of decentralization is better scalability. It is crucial that the system can expand rapidly from a few hundred peers to several thousand or even millions without deterioration of performance. In P2P systems, all peer machines provide a part of the service. Algorithms for resource discovery and search have to be capable of supporting the system's extensibility in taking into account resources shared by all participants in order to increase available resources.

- *Dynamism*: P2P systems must face intermittent participation of its nodes. Resources such as compute nodes and files, can join and leave the system frequently and unpredictably. In addition, the number of participant nodes is always in constant evolution. The P2P approach must be designed to adapt to such a highly volatile and dynamic environment.

- *Heterogeneity*: In P2P systems, supporting heterogeneity is needed because these systems are to be used by a great number of peers that do not belong to a common structure. Hence, it is impossible that these peers have an identical material architecture.

- *Fault resilience*: One of the primary design goals of a P2P system is to avoid a central point of failure. Although most pure P2P systems already achieve this goal, they have to face failures commonly associated with distributed systems spanning multiple hosts and networks, such as disconnection, unavailability, partitions, and node failures. It is necessary for the system to continue to operate with the still active peers in the presence of such failures.

- *Security*: It is crucial to protect the peer machines and the applications from malicious behavior that attempts to corrupt the operation of the system by taking control of the application.

4.3   Data location and routing algorithms

In P2P file sharing systems, each client shares some files and is interested in downloading some files from other peers. The location mechanisms and routing algorithms are crucial to the searching operations for a resource that the client wants. Queries to locate data items may be file identifiers or keywords with regular expressions. Peers are expected to process queries and produce results individually, and the total result set for a query is the bag union of results from every node that processes the query.

A P2P overlay network can be considered an undirected graph, where the nodes correspond to P2P nodes in the network, and the edges correspond to open connections maintained between the nodes. Two nodes maintaining an open connection between themselves are known as neighbors. Messages are transferred along the edges. For a message to travel from a source node to a destination node, it must travel along a path in the graph. The length of this traveled path is known as the number of hops taken by the message.

To search for a file, the user initiates a request message to other nodes; its node becomes the source of the message. The routing algorithm determines to how many neighbors, and to which neighbors, the message will be forwarded. Once the request message is received, the node will process the query over its local store. If the query is satisfied at that node, the node will send a response message back to the source of the message. In unstructured P2P networks,

Table 4.1: A comparison of different unstructured systems.

P2P system	Network structure
Napster	Hybrid decentralized system with central cluster of approximately 160 servers for all peers.
Gnutella	Purely decentralized system.
Freenet	Purely decentralized system. A loose DHT structure.
FastTrack/KaZaA	Partially centralized system.
eDonkey2000	Hybrid decentralized system with tens of servers around the world. Peers can host their own server.
BitTorrent	Hybrid decentralized system with central servers called *tracker*. Each file can be managed by a different tracker.

the address of the query source is unknown to the replying node. In this case, the replying node sends the response message along the reverse path traveled by the query message. In structured P2P networks, the replying node may know the address of the query source, and will transfer the response message to the source directly. In this section, we present several typical P2P routing algorithms in both unstructured and structured systems.

4.3.1 P2P evolution

First generation P2P systems consist of unstructured P2P systems such as Napster and Gnutella, which are basically easy to implement and do not contain much optimization. The broadcasting method used by unstructured systems for data lookup may have large routing costs or fail to find available content. Hence, more sophisticated systems based on Distributed Hash Table (DHT) routing algorithms are proposed and they are considered second generation P2P systems. Their purpose is when given a query to efficiently route a message to the destination. They create a form of virtual topology and are generally named *structured P2P systems*. Third generation P2P systems are also variants of structured P2P systems. However, they put more effort on security to close the gap between working implementations, security and anonymity.

4.3.2 Unstructured P2P systems

In unstructured systems, the overlay network is created in an ad-hoc fashion as nodes and content are added. The data location is not controlled by the system and no guarantees for the success of a search are offered to the users. In order to increase the probability of data lookup success, replicated copies of popular files are shared among peers. The core feature of widely deployed systems is file-sharing. Search techniques such as flooding, random walks, expanding-ring Time-To-Live (TTL) have been investigated in order to inquire at each peer in the system about the placement of the file. In [194], a list of algorithms used in unstructured P2P overlay networks is provided.

The main advantage of unstructured systems compared to the traditional client-server architecture is the high availability of files and network capacity among peers. Instead of downloading from the centralized server, peers can download the files directly from other peers in the network. Hence, the total network bandwidth for file transfers is far greater than any possible centralized server can provide.

We classify unstructured systems into three groups according to how the files and peer indexes are organized: *hybrid decentralized*, *purely decentralized* and *partially centralized*. The indexes which map files to peers are used to perform data lookup queries. In hybrid decentralized systems, peer indexes are stored on the centralized server. In purely decentralized systems, each peer stores its file indexes locally and all peers are treated equally. In partially centralized systems, different peers store their indexes at different super-peers. Figure 4.1 shows the classification of unstructured systems. In this section, we focus on data location issues and routing algorithms in unstructured P2P systems. We present some of the most popular unstructured systems: Napster/OpenNap, Gnutella, Freenet, FastTrack/KaZaA, eDonkey2000, BitTorrent.

4.3.2.1 Napster/OpenNap

Overview Napster is one of the most popular examples of file-sharing hybrid decentralized systems. Its protocol was not published but it was analyzed and there exists the OpenNap open source project[1] that follows the same specification [188]. Napster is considered the first unstructured P2P system that achieved global-scale deployment and was characterized as "the fastest growing Internet application ever" [180], reaching 50 millions users in just one year.

Routing algorithm for getting data Napster made popular the *centralized directory model* as the algorithm used for searching operations. An OpenNap server allows peers to connect to it and offers the same file lookup, browse capabilities offered by Napster. In order to make his files accessible to other users, a client has to send a list of files that he wants to share to a central directory server. The server updates its database and files are indexed by their name. In order to retrieve a file, a client sends requests for the file to the server about the list of peers storing the file. The user then chooses one or more of the peers in the list that hold the requested file and opens a direct communication with these peers to download it (see Figure 4.1).

Although only part of the protocol is based on client-server architecture, the system is considered P2P because only the file index is accessed in client-server mode and the digital objects are transferred directly between peers.

---

[1]http://opennap.sourceforge.net/

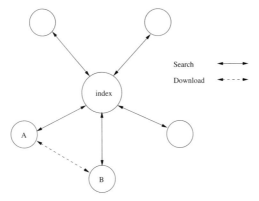

FIGURE 4.1: Typical hybrid decentralized P2P system. A central directory server maintains an index of files in the network.

Such systems with a central server are not easy to scale and the central index server used in Napster is a single point of failure.

4.3.2.2 Gnutella

Overview Gnutella[2] is a purely decentralized system with a flat topology of peers. A purely decentralized system is one that does not contain any central point of control or focus. Each node within the system is considered being of equal standing (i.e., servents).

To join the system, a new peer initially connects to other active peers that are already members of the system. There is a number of hosts well-known to the Gnutella community (e.g., list of hosts available from `http://gnutellahosts.com`) that can serve as an initial connection point. Once connected to the system, peers send messages to interact with each other. The Gnutella protocol supports following messages.

- Group membership (*PING* and *PONG*): a group membership message is either a Ping or a Pong. A peer joining the system broadcasts a Ping message to declare its own presence in the network. A Pong message, which contains information about the peer (e.g., IP address, number and size of the data item that it shares in the system) will be routed back along the opposite path through which the original Ping message arrived. In a dynamic environment like Gnutella where nodes often join and leave the network unpredictably, a node periodically pings its neighbors to discover other participating nodes. Using information from received pong messages, a disconnected node can always reconnect to the network.

---

[2]`http://gnutella.wego.com`

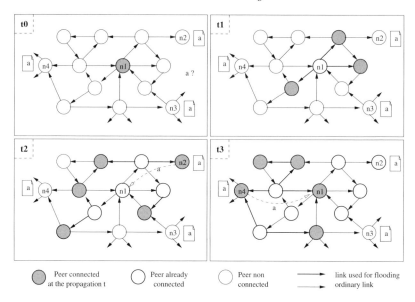

FIGURE 4.2: An example of data lookup in a flooding algorithm.

- Search (*QUERY* and *QUERY HIT*): a query message contains a specified search string and the minimum speed requirements of the responding peer. A peer possessing the requested resource replies with a QUERY HIT message that contains the information necessary to download a file (e.g., IP, port and speed of the responding host, the number of matching files found and their indexed result set).

- File transfer (*GET* and *PUSH*): file downloads are done directly between two peers using these types of messages.

Each peer in a Gnutella system maintains a small number of permanent links to neighbors (typically 4 or 5). In order to cope with the unreliability after joining the system, a peer periodically sends a Ping message to its neighbors to discover other participating peers.

Routing algorithm for getting data   The Gnutella algorithm uses the *flooded requests model* [183] for discovery and retrieval of data. In this model, to locate a data item peer **n** sends requests to its neighbors. Then, the requests will be flooded to their directly connected peers until the data item is found or a maximum number of flooding steps occur **b**, in the original protocol **b** = 7. Each peer, which receives the request, performs the following operations: (i) check for matches against their local data set; (ii) if yes, a notification is sent to **n**; (iii) otherwise if **b** > 0, the request is flooded to its neighbors in decrementing **b**.

Figure 4.2 illustrates an example of data lookup in the Gnutella network with b = 2 for the flooding step. Note that we use the directed graph not the undirected one to represent the overlay network in this example.

- At t0: peer n1 **n** searches for peers holding data item **a**.
- At t1: n1 sends a request to its neighbors, with b = 2. The request will be flooded two times. None of **n1**'s neighbors has data matching the request.
- At t2: n1's neighbors retransmit the request to its neighbors with b = 1. The peer **n2** holding the requested data **a** sends a notification to n1.
- At t3: last propagation of the request is performed with b = 0. Peer n4 notifies **n1** about its possession of data item **a**.

Once b reaches zero, the request is dropped. However, this algorithm does not ensure success of data lookup queries even if requested data items exist somewhere in the network, particularly when b is low (e.g., in the above example, peer **n3** holding the requested data item **a** is not contacted). If b is set higher in order to increase chance of finding data items, there will be many messages propagated even for only one query, particularly in high connectivity networks. Another problem of flooding is that it introduces duplicative messages, which are multiple copies of a query sent to a node by its multiple neighbors. These duplicative messages incur considerable extra load at the node receiving them and unnecessarily burden the network. Various solutions have been proposed to address the above issues (see Section 4.4.1).

4.3.2.3 Freenet

Overview Freenet is an example of a loosely structured decentralized system with the support of anonymity. Data items are identified by binary file keys, named Globally Unique Identifiers (GUID), obtained by applying the hash function (SHA-1) to the file name. The Freenet employs three kinds of GUIDs: *Keyword-Signed Key* (KSK) intended for human use, *Signed-Subspace Key* (SSK) like KSK but preventing namespace collisions, and *Content Hash Keys* (CHK) used for primary data storage.

KSK is the simplest type of GUIDs, which is derived from a descriptive text string chosen by the user. This descriptive text string is used to generate a public/private key pair whose public half is hashed to create the file key. The private half can be used to verify the integrity of the retrieved file. The file itself is encrypted using the user-defined descriptor as key. For finding the file, the user must know the descriptive text.

SSK prevents users from independently choosing the same descriptive string for different files. It also enables constructing personal namespaces, i.e., subspace, for example /text/poems/romantic/. SSK is composed of two parts. The first part is the public namespace key and the second part is the descriptive string chosen by the user. These two parts are hashed independently and concatenated together to be used as a search key. To retrieve a file from a subspace, the user needs only the subspace's public key and the descriptive string. Adding or updating a file requires the private key of subspace. Therefore, SSK facilitates trust by guaranteeing that updates of subspace are done only by its owner.

CHK is the low-level data-storage key, which is generated by computing the hash value of file content. CHK is useful for implementing updating and splitting of contents. A content-file hash guarantees that each file has a unique absolute identifier (SHA-1 collisions are considered nearly impossible). Since CHK keys are binary, they are not appropriate for user interaction. Hence, CHK keys usually are used in conjunction with SSK keys. A CHK key can be contained in an indirect file stored in user subspace. Given the SSK the original file is retrieved in 2 steps. First, the key value of actual filename is retrieved as the CHK key. Then, this CHK key can be used for a search in Freenet.

Routing algorithm for getting data   Freenet uses the *"Steepest Ascent Hill-Climbing Search"* algorithm. When a node receives a query, it first checks its own store. If the request is not satisfied, the node forwards the request to the node with the closest key to the one requested. If the chosen node cannot find the destination, it will return to the originator with a failed message. Otherwise, the node will try some other node. When the request is successful, each node which passed the request sends now the file back and creates an entry in its routing table binding the file holder with the requested key. On the way back the nodes might cache the file at their stores to improve search time for subsequent searches. However, to limit searches and resource usage, queries are forwarded within certain TTL values. Therefore, a node holding the requested data item will not be reachable if it is at the far end of the network.

4.3.2.4   FastTrack/KaZaA

Overview   FastTrack [155] is a partially centralized system that uses the concept of super-peers: peers with high bandwidth, disk space and processing power that are dynamically elected to facilitate searches by indexing shared files of a subpart of the peer network. KaZaA is a typical and widely used FastTrack application.

Routing algorithm for getting data   In a FastTrack network, a super-peer maintains an index of the files shared by peers connected to it. For example, in Figure 4.3 the peer n3 conserves information about data that its leaves n1 and n2 possess. All the queries initiated by ordinary peers are forwarded to the super-peer. Then, the search is performed using the *flooded requests model* in a highly pruned overlay network of super-peers. In comparison with purely decentralized systems, this approach has two major advantages: (i) the discovery time of files is reduced considerably, and (ii) the super-peers improve the network efficiency of the system by assuming a large portion of the entire

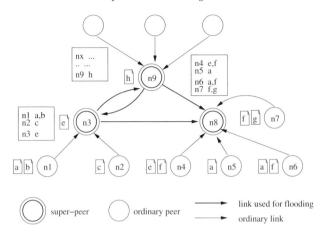

FIGURE 4.3: Peers and super-peers in partially centralized system.

network load. However, this approach still consumes bandwidth to maintain the index at the super-peers on behalf of the peers that are connected.

Typically, the number of leaves per super-peer is about sixty in FastTrack and about twenty in Gnutella v0,6. The number of peers assuming the data lookup is lower compared with Gnutella v0.4. The probability of finding data conforming to a query by flooding is higher, even though obtaining the data item is still not guaranteed more than with Gnutella v0.4.

4.3.2.5 eDonkey2000

Overview Overnet/eDonkey2000 [154], [153] is a hybrid two-layer decentralized system composed of client and server. There are loosely connected, separate index servers, but there is no single centralized server. These index servers are distributed over the world, unlike Napster which runs its central server in Silicon Valley. Although there are millions of clients, the number of servers is only several hundred. Moreover, the reliable servers that accept joining connections from new client nodes are only several dozen.

Routing algorithm for getting data With its protocol MFTP (Multisource File Transfer Protocol), eDonkey2000 allows download time to be optimized because the client peer can download concurrently parts of the file from multiple peers and during downloading it can share downloaded parts. File hashes are used to identify files on the network. This architecture uses *ed2k* link to store the metadata for clients who want to download a particular file. The link contains information about the file such as its name, size and the file hash or hash set (part hashes). The complete hash set ensures that blocks of the file are always correct and helps spreading new and rare files.

To join the network, a new peer needs to know the IP address and port of a bootstrapping peer (server) in the network. It then connects to this server and sends a list of all its shared files with the metadata describing these object files. Servers maintain a database with file object IDs mapped to server-generated client IDs. When a client wishes to download a file, it sends queries for this file to the directly connected server. The server looks at its database and returns a list of known sources. After having this list, it contacts the sources and asks to download the file.

4.3.2.6   BitTorrent

Overview   BitTorrent is a hybrid decentralized system that uses a centralized location to manage users' downloads. The BitTorrent protocol consists of five major components: (i) *.torrent files*, (ii) *a website*, (iii) *tracker server*, (iv) *client seeders*, and (v) *client leechers*.

Routing algorithm for getting data   A *.torrent* file is composed of a header and a number of SHA-1 block hashes of the original file. The header contains information about the file, its length, name and the IP address or URL of a tracker for this *.torrent* file. The *.torrent* file is stored on a public accessible website. The original content owner starts a BitTorrent client that possesses a complete copy of the file along with a copy of the *.torrent* file. Once the *.torrent* file is read, the BitTorrent client registers with the *tracker* as a *seeder* since it has a complete copy of the file. When a downloader gets the *.torrent* file from the public available website, the BitTorrent client then parses the header information as well as the SHA-1 hash blocks. Nevertheless, because it does not have a copy of the file, it registers itself with the tracker as a *leecher*. The tracker randomly generates a list of contact information about the peers that are downloading the same file and sends this list to the leecher. Leechers then use this information to connect to each other for downloading the file.

BitTorrent cuts files into pieces of fixed size (256 KB chunks) to track the content of each peer. The seeder and leechers transfer pieces of the file among each other using a *tit-for-tat* algorithm. This algorithm is designed to guarantee a high level of data exchange while discouraging free-riders: peers that do not contribute should not be able to achieve high download rates. When a peer finishes downloading a piece, the BitTorrent client matches that piece against the SHA-1 hash for that piece, which is included in the *.torrent* file. After data integrity of the piece is validated, that peer can announce to all of other peers that it has that piece for sharing. When a *leecher* has obtained all pieces of the file, it then becomes a pure *seeder* of the content. During the piece exchange process, a peer may join or leave the network. In order to avoid file exchange interruption, a peer re-requests an updated list of peers from the tracker periodically.

4.3.3   Structured P2P systems

In structured systems, the overlay network topology is tightly controlled and files are placed at precisely specified locations. These systems use a

Distributed Hash Table (DHT) to provide a mapping between the file identifier and location, so that queries can be efficiently routed to the node with the desired file.

Structured systems employ different DHT for routing messages and locating data. Some of the most interesting and representative DHT and their corresponding systems are examined in the following sections.

4.3.3.1 Overview of Distributed Hash Table

The goal of DHTs is to provide the efficient location of data items in a very large and dynamic distributed system without relying on any centralized infrastructure. A DHT applies the principle of a hash table: a data item has an identifier (e.g., in file system, the absolute path `/home/toto/book.pdf`). This identifier is sent to a hash function. Most of the time, the hash function is either SHA-1 [159] or MD5 [185], which generates with high probability a unique key in the same virtual space by hashing *key = hash(identifier)*. This pair of values (*identifier,key*) is completely one way, which means that having a similar hash value does not assume that the items are similar. Each peer in the system handles a portion of the hash space and is responsible for storing a certain range of keys. After a lookup for a certain key, the system will return the identity (e.g., the IP address) of the peer storing the data item with that key.

It is crucial that the hash function should balance the distribution of keys throughout the space. Peers should receive a random identifier in the DHT that evenly spreads in the space where each peer stores a similar number of data items. This ensures load balance of the system. The size of the hash function's output space must be large enough so that the probability of key collision between two different data items is minimized. The division of key address space varies between systems. Some organize into some shapes like rings, trees and hypercube. A guarantee, which is usually logarithmic, is given that the final destination will finally be reached in several steps.

Next, we review some of the most important DHTs.

4.3.3.2 Chord

Overview In Chord [193], peers are uniformly distributed in a logical ring ordered by increasing order of identifier, which is called an *identifier circle* or *Chord ring*. The identifiers are determined by means of a deterministic function, a variant of consistent hashing [172]. Consistent hashing is designed to balance the load on the system, since each peer receives roughly the same number of keys and there is a minimal impact on the movement of keys when peers join or leave the system. A peer's identifier is chosen by hashing the peer's IP address, while a key identifier is produced by hashing the data key. Identifiers are ordered in the ring according to the modulo of the key with the number 2^m. Suppose that the key consists of m bits. Key k is assigned to

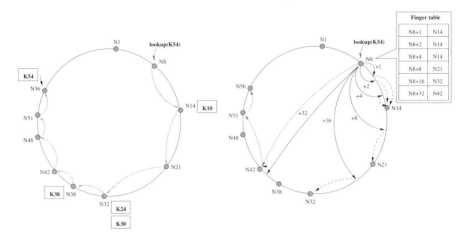

FIGURE 4.4: Chord ring with identifier circle consisting of ten peers and five data keys. It shows the path followed by a query originated at peer N8 for the lookup of key 54. Finger table entries for peer N8.

the first peer whose identifier is equal to, or follows k in the identifier space. This node is called the successor node of key k.

Routing algorithm for getting data   Each peer in the Chord ring needs to know how to contact its successor peer on the circle for routing messages. Queries for a given identifier k are passed around the circle via the successor pointers until they first encounter a node that includes the desired identifier. This is the node the query maps to. This simple key lookup is shown in Algorithm 4.3.1.

Algorithm 4.3.1 Simple key lookup using the finger table.

Function find_successor(k)
1: **if** $id \in (n, successor]$ **then**
2:     *return* successor;
3: **else**
4:     *// forward the query around the circle*;
5:     *return* successor.find_successor(k);
6: **end if**

However, in the worst case, queries need to traverse all peers to find a certain key. In order to improve routing performance, each Chord peer maintains a routing table with up to m entries, called *finger tables*, where 2^m is the number of possible identifiers. The first entry points to its immediate successor on the circle. The ith entry in the table at peer n contains the identity of the first peer s that succeeds n by at least 2^{i-1} on the identifier circle (i.e., s = successor$(n+2^{i-1})$, where $1<i<m$).

The algorithm for scalable key lookup is presented in Algorithm 4.3.2. For a node n to perform a lookup for key k, the finger table is consulted to identify the first largest peer n' whose identifier is between n and k. If such a peer exists, *find_successor* is done and node n returns its successor n' and the lookup is repeated starting from n'. The procedure continues until the peer that stores the key is found. Otherwise, n searches its finger table for the node n' whose identifier most immediately precedes k, and then invokes *find_successor* at n'. The reason behind this choice of n' is that the closer n' is to k, the more it will know about the identifer circle in the region of k. In a system with N peers, when a peer executes a lookup operation, $O(logN)$ messages are transmitted to other peers.

An example of a Chord ring and the finger table entries of the peer N8 is shown in Figure 4.4. The rows of the first column result from the computation $n+2^{i-1}$; the second column is the successor of this identifier. In the example, peer N8 executes a lookup operation for the data key 54, and it has to visit all peers between peer N8 and peer N56 before the data key 54 is found. The lookup complexity is $O(N)$. With the finger table, the sequence of hops is reduced considerably (e.g., N42/N51/N56). The lookup complexity is $O(logN)$.

Algorithm 4.3.2 Scalable key lookup using the finger table.

Function find_successor(k)

1: **if** $id \in (n,successor]$ **then**
2:     *return* successor;
3: **else**
4:     n' = closest_preceding_node(k);
5:     *return* n'.find_successor(k);
6: **end if**

Function closest_preceding_node(k)

1: **for** $i = m$ downto *1* **do**
2:     **if** *finger[i]* $\in$ *(n,k)* **then**
3:       *return* finger[i];
4:     **end if**
5: **end for**
6: *return* n;

Handling when nodes enter and leave When a peer n enters the system, it uses the information from the predecessor and successor of itself to initialize, store keys and configure pointers. The joining peer begins by hashing its IP address (or takes a key randomly) to have a key determine its position in the Chord ring. Then, n sends a message to the peer n' holding this key. This peer initializes the finger table of node n by delegating to n keys previously assigned to it. The peer n' becomes also the successor of n. The predecessors of n' are notified about its new successor n to update its fingers to reflect the change in the network topology caused by addition of n. Finally, all keys for which n has become successor are transferred to n. The node joining algorithm is presented in Algorithm 4.3.3 [192].

Similarly, when peer n leaves the Chord system, all of its assigned keys are reassigned to n's successor to maintain consistent hashing mapping. No other changes of keys assignment to peers need to take place. It is crucial for the correctness of the Chord protocol that each peer is aware of its successors. When peers fail, it is possible that a peer does not know its new successor and it has no chance to learn about it. To avoid this situation, a peer maintains a successor list of size r. A peer simply contacts the next peer on its successor list when its immediate successor peer does not respond. The probability that all r peers in the successor list fail is p^r where p is the probability for a peer failure.

Chord has been widely used in research and it gives certain guarantees on finding a data item. In Chord, the number of hops traversed for accessing node A from node B might be different from traversing from node B to node A. This leads to the link asymmetry problem. The other drawback of Chord is node joining cost. The distribution neighbors and number of entries in the routing table depend mostly on the location of the node on the circle. A new joining node needs to contact its immediate successor, which can be at the other side of the network to initiate first connection. This is expensive in terms of maintaining consistency and network bandwidth.

4.3.3.3 Content-Addressable Network (CAN)

Overview CAN is a distributed hash-based architecture that maps file names to their location in the network. In this work, authors are seeking an Internet-scale naming system, which is location-independent and fault-tolerant. Each node of the CAN network stores a chunk (referred to as a "zone") of the entire hash table, as well as information about its neighbors. Requests to insert, lookup or delete a particular key are routed to those nodes whose zone have the corresponding keys.

Routing algorithm for getting data The CAN [182] design centers around a virtual d-dimensional Cartesian coordinate space to store (*key k, value v*) pairs. This coordinated space is partitioned into segments corresponding with zones in the hash table. These zones are distributed to all the nodes of the system. In that way, each node takes responsibility of a zone of the hash table. For example, Figure 4.5 shows a 2-dimensional [0,1] x [0,1] coordinate space partitioned among 5 nodes. Any key k is mapped onto a point p using a hash function on the address space. Then, the corresponding (k,v) pair is stored to the node whose zone includes point p. For example, in the case of Figure 4.5, a key that maps to coordinate (0.1,0.2) would be stored at the node responsible for zone A.

Algorithm 4.3.3 Node joining algorithm in Chord.

Function join(n')

1: // Node n join the Chord network, node n' is an arbitrary node in the network
2: **if** *n'* **then**
3:     init_finger_table(n');
4:     notify();
5:     s = successor; // get successor
6:     s.move_keys(n);
7: **else**
8:     // no other node in the network to n itself
9:     **for** *i = 1 to m* **do**
10:         finger[i].node = n;
11:     **end for**
12:     predecessor = successor = n;
13: **end if**

Function init_finger_table(n')

1: // initialize finger table of local node, n' is an arbitrary node already in the network
2: finger[i].node = n'.find_successor(finger[1].start);
3: successor = finger[1].node;
4: **for** *i = 1 to (m - 1)* **do**
5:     **if** *finger[i+1].start ∈ [n,finger[i].node)* **then**
6:         finger[i+1].node = finger[i].node;
7:     **else**
8:         finger[i+1].node = n'.find_successor(finger[i+1].start);
9:     **end if**
10: **end for**

Function notify()

1: // update finger tables of all nodes for which local node, n, has became their finger
2: **for** *i = 1 to m* **do**
3:     // find closest node p whose i^{th} finger can be n
4:     p = find_predecessor(n - 2^{i-1});
5:     p.update_finger_table(n,i);
6: **end for**

Function update_finger_table(s,i)

1: // if s is i^{th} finger of n, update n's finger table with s
2: **if** *s ∈ [n,finger[i].node)* **then**
3:     finger[i].node = s;
4:     p = predecessor; // get first node preceding n
5:     p.update_finger_table(s,i);
6: **end if**
7: finger[next] = find_successor(n + 2^{next-1});

Function move_keys(p)

1: // if p is new successor of local stored key k, move k (and its value) to p
2: **for** *each key k stored locally* **do**
3:     **if** *p ∈ [d,n)* **then**
4:         move k to p;
5:     **end if**
6: **end for**

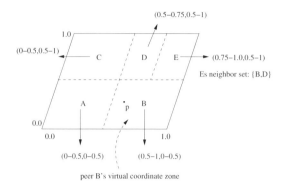

FIGURE 4.5: Example of a 2-d space with 5 nodes.

To retrieve the key k and the corresponding data, a node executes the same hash function to find point p and then retrieve the corresponding value v from the node covering p. For the routing operation, each node maintains the IP address of its neighbors having zones adjoining to its own and constructs a coordinate routing table. The request is routed from node-to-node until it reaches the node covering p. CAN uses a greedy algorithm to route messages (see the Algorithm 4.3.4), where each node sends the message to the neighbor that is closest to destination. Figure 4.5 depicts an example that illustrates a routing process; a request from node C for a key mapping to point p would be routed through node A to point p along the straight line represented by the arrow. A node tries to reach a destination using the best candidate from the routing table. However, in case it fails, the routing table will backtrack and choose another candidate. If connections to all neighbors are lost, the Expanding Ring Search (ERS) algorithm will be used to discover any node that is connected to the network.

Algorithm 4.3.4 Greedy algorithm for routing messages in CAN.

Function search(from_node, to_node, count)

1: **if** *from_node* $=$ *to_node* **then**
2:    // *search successful*
3:    *return* from_node;
4: **end if**
5: **if** *count* $=$ *network_size* **then**
6:    // *search failed*
7:    *return* null;
8: **end if**
9: visited[from_node] := true
10: **for** *each neighbor of from_node* **do**
11:    Find the distance of neighbors from from_node
12: **end for**
13: next_node := closest neighbor of from_node;
14: search(next_node, to_node, count+1)

Handling when nodes enter and leave Algorithm 4.3.5 is used to handle node joining in CAN. When a new node joins the system, it is allocated a portion of the coordinate space by splitting the zone of an existing node in half. Concretely, this operation is performed in three steps. Firstly, the new node discovers a bootstrapping node in the CAN network which maintains a list of active CAN nodes. Secondly, the node connects to the bootstrapping node and the new joining node randomly chooses a point p in the coordinate space and sends a *JOIN* request to the node covering p. The zone is split and a half is assigned to the new node. Finally, keys that belong to the partitioned zone will be transferred to the new node. Additionally, the neighbors of the split zone are notified to update their routing table to include the new node.

Algorithm 4.3.5 Node joining algorithm in CAN.

1: Add first node to the graph.
2: Choose a random position in the virtual coordinate space for new_node.
3: Add new_node to the graph.
4: Find the old_node that owns that position.
5: Split the region occupied by old_node along the longer axis and assign one half to the new_node.
6: Fix the edges in the graph for both the old_node and new_node.
7: nodes_added = nodes_added + 1.
8: Repeat Step 2 if nodes_added < desired_network_size.

When a node leaves the system, its zone and its hash table entries are assigned to one of its neighbors. Periodically, a node sends update messages to each of its neighbors about its zone coordinates, its neighbor list. A node detects failure of other nodes if it does not receive such update messages within a certain amount of time and initiates a controlled takeover mechanism. However, it will be listening for takeover messages from the surroundings. If any other node finishes the takeover before the current node, takeover process is canceled. The takeover is done to recombine address space.

The major advantage of CAN is in the fact that it performs better in terms of node states. A CAN node only keeps $2d$ states with d, dimensional space, while a node in other systems usually has around *logN* states. However, some drawbacks of CAN are mainly present in the CAN routing algorithm, which is not as efficient as the others. To increase routing performance, authors propose to use Round Trip Timer (RTT)-weighted routing, replication of entries and increasing dimensions. Though increasing dimensions leads to the cost of increased per-node state, it reduces considerably application-level hop count. Since each node will have multiple neighbors connected to each other, contacting a next hop will take smaller paths.

4.3.4 Hybrid P2P systems

Both unstructured and structured systems have advantages and disadvantages. Unstructured systems have low cost for maintenance of network structure. However, query processing in these systems is not very efficient and does not scale well. These drawbacks arise because they create a random graph that represents the network topology, where queries are propagated from node to node in a blind manner. On the other hand, data files in structured systems are placed not at random nodes but at specified locations using hash functions. Such systems have good performance for point queries, but they are not efficient for text or range queries.

Several hybrid systems have been proposed to overcome the drawbacks of each while retaining their benefits. In these systems, a peer's neighbor connections are defined more flexibly than those in structured systems.

4.3.4.1   Pastry

Overview   Pastry [186] is generic peer-to-peer content location very similar to Tapestry, proposed by Rice university in cooperation with Microsoft Research Center. Pastry differs from other P2P routing substrates such as Chord and CAN mainly in its approach for achieving network locality and object replication.

In Pastry, messages are routed to nodes based on the provided keys. A Java version of Pastry has been implemented: FreePastry. Based on this routing layer, other applications such as Scribe [187], PAST [158] and Squirrel [169] have been developed.

Routing algorithm for getting data   Each Pastry node has a unique, 128-bit identifier called the *nodeID* that is produced from computing a cryptographic hash of the node's public key or its IP address. This procedure guarantees unique nodeIDs. The uniform distribution of NodeIDs ensures an even population of the nodeID space. Each data also has a 128-bit key. This key is generated by a hash function. The data is stored in the node whose nodeID is numerically closest to its key.

Each node divides its routing table in three parts: routing table, neighborhood set and leaf set. The first part is the "routing table", which includes information on peers needed to route messages according to the description made before. Assuming a network consisting of N nodes, a node's routing table is organized into *logN* rows with *2b-1* entries each row. Note that b is a configuration parameter of the Pastry system with typical value 4. The nth row of the routing table contains the nodeIDs and IP addresses of those nodes, whose nodeID shares the present node's nodeID in the first n digits but different in the *n+1* digit. If there are more than *2b-1* qualified nodes, the closest *2b-1* nodes will be selected, according to a proximity metric such as the delay or the number of IP routing hops. The "leaf set" allows nodes to know exactly which key belongs to them and which keys belong to their neighbors. It contains the nodeIDs and IP addresses of the half nodes with numerically closest larger nodeIDs, and half nodes with numerically closest smaller nodeIDs, relative to the present node's nodeID. Finally, the

Table 4.2: Notation definition for Algorithm 4.3.6.

Notation	Comment
R_l^i	The entry in the routing table R at column i, $0 \leq i < 2^b$ and row l, $0 \leq l < 128/b$
L_i	The i-th closest nodeID in the leaf set L, $-\lfloor|L|/2\rfloor \leq i \leq \lfloor|L|/2\rfloor$, where negative/positive indexes indicate nodeIDs smaller/larger than the present nodeID, respectively.
D_l	The value of the l's digit in the key D
$shl(A,B)$	The length of the prefix shared among A and B, in digits

"neighborhood set" is a list of nodes that contains the nodeIDs and IP addresses that are physically closest to the present node.

Algorithm 4.3.6 Routing algorithm in Pastry.

1: **if** $L_{-\lfloor|L|/2\rfloor} \leqslant D \leqslant L_{\lfloor|L|/2\rfloor}$ **then**
2:     // D is within range of our leaf set
3:     forward to L_i, so that $|D - L_i|$ is minimal;
4: **else**
5:     // use the routing table
6:     Let l = shl(D, A);
7:     **if** $R \neq null$ **then**
8:         forward to $R_l^{D_l}$;
9:     **else**
10:         // rare case
11:         forward to T $\in$ L $\cup$ R $\cup$ M, so that
12:         shl(T, D) $\geq$ l,
13:         |T - D| < |A - D|
14:     **end if**
15: **end if**

For message routing, the node tries to route client messages to the node with a nodeId that is numerically closest to the key, among all live Pastry nodes. In each routing step, the message reaches a node sharing a prefix (with the target object) of one digit longer, thus reaching the destination in $O(logN)$ hops [186] where N is the number of nodes in the system.

Algorithm 4.3.6 which uses notations in Table 4.2 is applied to route messages in Pastry. If the node finds that the key falls within the range of nodeIDs covered by its leaf set (line 1), it directly forwards the message to the node in the leaf set whose nodeID is closest to the key (line 3). If the key is not covered by the leaf set, then the routing table is used and the message is forwarded to a node that shares a common prefix with the key by at least one more digit (line 6-8). In certain cases, it is possible that the appropriate entry in the routing table is empty or the associated node is not reachable (line 11-14), in which case the message is forwarded to a node that shares a

Table 4.3: State of a Pastry node with node ID 23002, b = 2. The top row of the routing table represents level zero. The neighborhood set is not used in routing, but is needed during node addition/recovery.

NodeID 23002			
Leaf Set			
Smaller		Greater	
23001	22333	23022	23033
22321	22312	23100	23101
Routing Table			
-0-1023	-1-2131	2	-3-0231
2-0-021		2-2-032	3
0		23-2-33	
0		230-2-2	
		2	
Neighborhood Set			
02132	32100	00213	10023
31102	22311	02310	01213

prefix with the key at least as long as the present node, and is numerically closer to the key than the present node's nodeId. Such a node must be in the leaf set unless the message has already arrived at the node with numerically closest nodeID.

Handling when nodes enter and leave   Table 4.3 [186] shows the state of a Pastry node with the nodeID 23002, b = 2 in a system that uses 5 digits for the node identifier. Node identifiers are split in three parts: equal prefix, current digit and different suffix. The first row keeps addresses of nodes that have no common prefix with current node. The second row keeps addresses of nodes that share the first digit with the current node and so on. At each row, the cell whose digit matches the node's digit has a gray background.

The authors claim that Pastry is efficient in terms of routing table size with a configuration parameter b of 4 and a node number of 10^6; a routing table contains about 75 entries and the expected number of hops to reach a destination is around 5. As each node maintains in its routing table state information of the network, there is a need to keep consistency. The authors propose exchanging periodical messages between nodes in the close address space. If a node does not respond for a certain amount of time, all the nodes update their routing table. Nevertheless, such exchanging of update messages wastes the network bandwidth.

4.3.4.2   Tapestry

Overview   Tapestry [197] is designed as a routing and location layer in OceanStore [173]. If Chord and CAN rely simply on hop count, which can sometimes take the lookup to the other side of the network, Tapestry considers the network distance when looking up keys. Tapestry provides an environ-

Peer-to-peer data management 119

Table 4.4: The neighbor map held by Tapestry node with ID 67493.

	Level 5	Level 4	Level 3	Level 2	Level 1
Entry 0	07493	x0493	xx093	xxx03	xxxx0
Entry 1	17493	x1493	xx193	xxx13	xxxx1
Entry 2	27493	x2493	xx293	xxx23	xxxx2
Entry 3	37493	x3493	xx393	xxx33	xxxx3
Entry 4	47493	x4493	xx493	xxx43	xxxx4
Entry 5	57493	x5493	xx593	xxx53	xxxx5
Entry 6	*67493*	x6493	xx693	xxx63	xxxx6
Entry 7	77493	x7493	xx793	xxx73	xxxx7
Entry 8	87493	x8493	xx893	xxx83	xxxx8
Entry 9	97493	x9493	xx993	xxx93	xxxx9

ment that offers system-wide stability, transparently masking faulty components, bypassing failed routes, removing nodes under attack from service and rapidly adapting communication topologies to circumstances. Location information is used for incrementally forwarding messages from point to point until they reach their destination. The consistency of location information is reparable on the fly, and if lost due to failures or destroyed, it is easily rebuilt or refreshed.

Routing algorithm for getting data Tapestry mechanisms are modeled after the Plaxton *mesh* [181]. The Plaxton data structure allows messages to locate objects and route to them across an arbitrarily-sized network, while using a small constant-sized routing map at each hop. In Plaxton, each node or machine can take on the roles of *servers* (where objects are stored), *routers* (which forward messages), and *clients* (origins of requests).

In the Plaxton mesh, each node has a *neighbor map* with constant size as shown in the example in Table 4.4. In a system with N-sized namespace using identifiers of base b, the neighbor map size is $b \, log_b(N)$. The neighbor map is organized into routing levels, and each level l contains a number of entries that point to a set of nodes matching the suffix for that level with l digits. The ith entry in the jth level is the identifier and location of the closest node which ends in "i" + suffix(N,j-1). For example, the 5th entry for the 3rd level for node 67493 points to the node closest to 67493 in network distance whose ID ends in 593.

Tapestry performs message routing as follow. Consider that some node S = 67493 sends a message to node $D = 34567$. To succeed, S has to discover some node that ends with a 7. Consider it to be $R1 = 98747$. Now, $R1$ must know about some node that ends with a 7 and has 6 in its former position, say $R2 = 64267$. This reasoning goes on for nodes $R3 = 45567$, $R4 = 64567$ and finally $D = 34567$. So the digits are resolved right to left as follows: xxxx7 → xxx67 → xx567 → x4567 → 34567.

Handling when nodes enter and leave The Plaxton mesh does not support dynamic node insertion and deletion, and does not handle node failure.

In other words, it supposes to be a static data structure. Tapestry extends its design to adapt it to the transient populations of peer-to-peer networks. It employs an incremental algorithm for node insertion, populating neighbor maps and notifying neighbors of new node insertions. Firstly, the joining node N requests a new identifier *new_id*. Then, it contacts the gateway node G, a Tapestry node known to N that acts as a bridge to the network. Starting with node G, node N attempts to route to *new_id*, and copies an approximate neighbor map from the ith hop H_i, $G = H_0$. Then, the relevant nodes are informed to take into account the joining node in their neighbor maps.

In order to leave the network, a node broadcasts its intention of leaving and transmits the replacement node for each level in the routing tables of the other nodes. Objects at the leaving node are redistributed or replenished from redundant copies.

4.4   Shortcomings and improvements of P2P systems

4.4.1   Unstructured P2P systems

The main characteristic of unstructured systems is that the peers are organized into a random graph and the placement of data is completely unrelated to the overlay network. As a result, a peer has no idea about which peers hold the relevant files that it desires. Data lookup in such systems is essentially based on the flooding method. Therefore, peers are placed under a high load handling distributed searches. Flooding is not scalable since it produces a large volume of unnecessary traffic in the network. However, according to [178], apart from scalability concerns unstructured systems might be the preferred choice for file-sharing and other applications as they offer the following advantages:

- Keyword searching is the common operation.

- Most content is typically replicated at a fair fraction of participating sites.

- The node population is highly transient.

- Users will accept a best-effort content retrieval approach.

A number of initiatives to address the scalability issue in unstructured systems have been initiated in recent years. Several solutions have been proposed to improve the scalability for unstructured systems.

The *random walk* method [176] has been proposed to replace the original flooding approach. In this method, each peer chooses an equal number of neighbors at random, and propagates its requests only to them. In this case, the TTL parameter designates the number of hops the walker should propagate. The use of random walks is found to significantly improve the performance of the system as the messages are reduced considerably. However, the drawback of this algorithm is its highly variable performance. Choice made of random peers has a great impact on success rates of data lookup. Another disadvantage of this method is its inability to adapt to different query loads. Queries for popular and unpopular objects are treated in the exact same manner without previous successes or failures are analyzing. As an alternative, *multiple parallel random walks* is proposed in [177]. Although compared to the single random walk this method has better behavior, it still suffers from low network coverage and slow response time. Hybrid methods that combine flooding and random walks have been proposed in [164].

In [196], authors suggest the use of a more sophisticated technique called *directed breadth first search (direct BFS)*. The choice of peers to propagate messages is based on their past history and local indexes, which are data structured where each node maintains very simple and small indexes over other nodes' data. Direct BFS makes use of local indexes by forwarding requests only to those peers that have often provided results to past requests, under the assumption that they will continue to do so. Then, the requests are iteratively forwarded to more nodes at increasing depths until the query is answered (i.e., *iterative deepening* technique).

In [171], a *localized search* mechanism is proposed where each node maintains an index or a profile of its neighbors' content that is used to rank its neighbors. Queries are forwarded selectively to the most appropriate profiles only and search is then restricted to what are believed to be neighbors with relevant results.

In [157], the use of *routing indexes* were introduced to address the searching and scalability issues where various types of indexes were proposed based on the way each index takes into account the information about the number of hops required for locating a matching peer.

In another family of algorithms, [195] and [156] take into account the different connectivity and forwarding capacities of peers in unstructured P2P systems to improve their scalability and search efficiency. Experiments demonstrated that peers with low bandwidth connections (i.e., nodes connected over dial-up modems) are easily saturated by flooding requests and thus slow down resource discovery in unstructured P2P systems. Hence, in order to improve system performance, low bandwidth peers need to be isolated from query routing. In these systems, peers with different bandwidth connections are distinguished into a two-level hierarchy of peers. High bandwidth peers, known as *super-peers*, form an unstructured overlay network, while peers with low bandwidth, the *leaves*, are connected only to super-peers. Each super-peer has an index of all the files contained in its leaves. Any request originating at a leaf peer is forwarded through the super-peers it is connected to, while flooding is performed only at the super-peer overlay network. This modification allows the system to retain the simplicity of unstructured systems while offering improved scalability.

In [184], the connectivity and reliability of unstructured networks (and in particular Gnutella) is studied. P2P networks such as Gnutella exhibit the properties of so-called power-law networks, in which the number of nodes with L links is proportional to L^{-k}, where k is a network-dependent constant. In other words, most nodes have few links, thus a large fraction of them can be taken away without seriously damaging the network connectivity, while there are a few highly connected nodes, which, if taken away, will cause the whole network to be broken down in pieces. One implication of this is that such networks are robust when facing random node attacks, however vulnerable to well-planned attacks.

4.4.2 Structured and hybrid P2P systems

Flooding search methods seem to be inherently scalable. As a result, search methods using distributed routing tables are proposed in structured systems. Concretely, these search methods are based on distributed hash tables (DHT), which provide a hash table interface with primitive methods *put(key,value)* and *get(key)*. We have introduced in this chapter four typical structured systems: Chord, CAN, Pastry, and Tapestry. In these systems, each peer is responsible for storing the values (file contents) corresponding to a certain range of keys. Most of the routing algorithms used for data lookup are equivalent in terms of routing table space cost which is *O(logN)* where N is the number of peers in the network. Similarly, the performance of structured systems is mostly *O(logN)*, with the exception of CAN, where the performance is given by $O(\frac{d}{4}N^{\frac{1}{d}})$, d being the number of employed dimensions.

The advantage of DHT-based searches is that queries can be efficiently routed to the nodes possessing the desired files. Since overlay network structure is strictly controlled, node joining or leaving operations in structured systems are mostly costly, in particularly for Chord, as node joining or leaving induces change to all other nodes.

However, one of the major limitations of structured P2P systems is that queries are typically limited to "exact-match" keyword search (as opposed to keyword queries). An exact identifier (key) of a data item should be appointed in order to locate the nodes that store that item. In practice, P2P users tend to submit queries with partial information for searching data items (e.g., all the songs by "Bryan Adams"). The support of searching based on multiple keywords is hence desirable. Active research is ongoing to extend the capabilities of structured systems to deal with more general queries such as range queries and join queries [168]. Nevertheless, it is arguable whether these capabilities can be efficiently implemented in a large-scale network.

In [170], a hybrid structured network is proposed where nodes organized in a structured network form its backbone. Each node in the backbone is also the leader of a cluster formed by non-backbone nodes. Within a cluster, nodes form an unstructured network and cooperate to store data and answer

Table 4.5: A comparison of various unstructured P2P systems.

Algorithm Taxonomy	Unstructured P2P system					
	Napster	Gnutella	Freenet	FastTrack/ KaZaA	eDonkey2000	BitTorrent
Decentralization	No explicit central server, peers are connected to central index server to locate data	Topology is flat with equal peers	Loosely DHT functionality	No explicit central server, peers are connected to super-peers	Hybrid two-layer network composed of clients and servers	Centralized model with a Tracker keeping track of peers
Architecture	Two-level hierarchical network of central index servers and peers	Flat and *ad-hoc* network of servants (peers). Flooding request and peers download directly	Keywords and descriptive text strings to identify data objects	Two-level hierarchical network of super-peers and peers	Servers provide the locations of files to requesting clients for download directly	Peers request information from a central Tracker
Lookup protocol	Central index server	Query flooding	Keys, Descriptive Text String search from peer to peer	Super-peers	Client-Server peers	Tracker
System Parameters	None	None	None	None	None	*.torrent* file
Routing Performance	Guarantee to locate data using central index servers	No guarantee to locate data. Improvements made in adapting Ultrapeer-client topologies. Good performance for popular content	Guarantee to locate data using *Key* search until the requests exceeded the Hops-To-Live (HTL)limits	Some degree of guarantee to locate data, since queries are rooted to super-peers which has a better scaling. Good performance for popular content	Guarantee to locate data and guarantee performance for popular content	Guarantee to locate data and guarantee performance for popular content
Routing State	Constant	Constant	Constant	Constant	Constant	Constant but choking (temporary refusal to upload) may occur
Peer Join/Leave	Constant	Constant	Constant	Constant	Constant with bootstrapping from other peers and connect to server to register files being shared	Constant
Security	Low. Threats: vulnerable to censorship, legal action, surveillance, malicious attack, and technical failure	Low. Threats: flooding, malicious content, virus spreading, attack on queries, and denial of service attacks	Low. Suffers from man-in-middle and Trojan attacks	Low. Threats: flooding, malicious or fake content, viruses, etc. Spywares monitor the activities of peers in the background	Moderate. Similar threats as the Fast-Track and BitTorrent	Moderate. Centralized Tracker manage file transfer and allows more control which makes it much harder faking IP addresses, port numbers, etc.
Reliability/Fault Resiliency	Degradation of the performance as there are bound to be limitations to the size of the server database and its capacity to respond to queries. Central point of failure	Degradation of the performance. Peers receive multiple copies of replies from peers that have the data. Requester peer can retry	No hierarchy or central point of failure exists	The ordinary peers are reassigned to other super-peers	Reconnecting to another server. Will not receive multiple replies from peers with available data	The Tracker keeps track of the peers and availability of the pieces of files.

Table 4.6: A comparison of various structured P2P systems.

Algorithm Taxonomy	Structured P2P system			
	Chord	CAN	Pastry	Tapestry
Decentralization	DHT functionality on Internet-like scale			
Architecture	Ini-directional and circular NodeID space	Multi-dimensional ID coordinate space	Plaxton-style global mesh network	Plaxton-style global mesh network
Lookup protocol	Matching Key and NodeID	$key, value$ pairs to map a point P in the co-ordinate space using uniform hash function	Matching Key and prefix in NodeID	Matching suffix in NodeID
System Parameters	N-number of peers in network	N-number of peers in network d-number of dimensions	N-number of peers in network, b-number of bits, $(B=2^b)$ used for the base of the chosen identifier	N-number of peers in network, B-base of the chosen identifier
Routing Performance	$O(logN)$	$O(d.N^{\frac{1}{d}})$	$O(log_B N)$	$O(log_B N)$
Routing State	$logN$	$2d$	$Blog_B N + Blog_B N$	$log_B N$
Peer Join/Leave	$(logN)^2$	$2d$	$log_B N$	$log_B N$
Security	Low level. Suffers from man-in-middle and Trojan attacks			
Reliability/Fault Resiliency	Failure of peers will not cause network-wide failure. Replicate data on multiple consecutive peers. On failures, application retries	Failure of peers will not cause network-wide failure. Multiple peers responsible for each data item. On failures, application retries	Failure of peers will not cause network-wide failure. Replicate data across multiple peers. Keep track of multiple paths to each peer	Failure of peers will not cause network-wide failure. Replicate data across multiple peers. Keep track of multiple paths to each peer

queries. When inserting a data item, multiple copies of its index are stored in a few different clusters. A query is also mapped to multiple clusters, and a flooding search within these clusters is performed. The union of all the search results are returned to users as the final result.

In [163], authors propose an approach that relies on multiple indexes, organized hierarchically, which permit users to locate data even using scarce information, although at the price of a higher lookup cost. The data itself is stored on only one (or a few) of the nodes.

In [165], an approach based on an underlying peer-to-peer system for both indexing and routing, and implementing a parallel query processing layer on top of it, is proposed.

It is argued in [167] that by creating keys for accessing data items (i.e., "virtualizing" the names of the data items) two main problems arise:

- *Locality is destroyed* Data items (i.e., files) from a single site are not usually co-located, meaning that opportunities for enhanced browsing, pre-fetching and efficient searching are lost.

- *Useful application level information is lost* The data used by many applications is naturally described using hierarchies, which expose relationships between items near to each other. The virtualization of the file namespace by generating keys discards this information.

How to maintain the overlay structure for routing algorithms to function efficiently in a very transient environment where nodes are joining and leaving at a high rate is an open research issue. The resilience of structured P2P systems in the face of a very transient user population is considered in [174].

4.5 Concluding remarks

This chapter presents an overview of P2P systems and the underlying characteristics of them. We discuss then certain unstructured, structured, and hybrid systems including routing algorithms for data lookup in each type of system. A table of characteristics, which summarizes the shortcomings and improvements of these systems is also provided.

References

[152] Peer-to-peer working group.

[153] Overnet/eDonkey2000, 2000. Available online at: `http://www.edonkey2000.com` (Accessed August 31st, 2007).

[154] The overnet file-sharing network, 2002. Available online at: `http://www.overnet.com` (Accessed August 31st, 2007).

[155] The FastTrack web site, 2003. Available online at: `http://www.fasttrack.nu` (Accessed August 31st, 2007).

[156] Y. Chawathe, S. Ratnasamy, L. Breslau, N. Lanham, and S. Shenker. Making gnutella-like P2P systems scalable. In *Proceedings of the 2003 Conference on Applications, Technologies, Architectures, and Protocols for Computer Communications (SIGCOMM '03)*, pages 407–418, New York, NY, 2003. ACM Press.

[157] A. Crespo and H. Garcia-Molina. Routing indices for peer-to-peer systems. In *Proceedings of the 22nd International Conference on Distributed Computing Systems (ICDCS'02)*, 2002.

[158] P. Druschel and A. I. T. Rowstron. PAST: A large-scale, persistent peer-to-peer storage utility. In *HotOS*, pages 75–80, 2001.

[159] D. Eastlake and P. Jones. US secure hash algorithm 1 (SHA1), 2001.

[160] S. Fanning. Napster home page, 2001. Available online at: `http://www.napster.com` (Accessed August 31st, 2007).

[161] I. Foster and A. Iamnitchi. On death, taxes, and the convergence of peer-to-peer and grid computing. In *Proceedings of 2nd International Workshop on Peer-to-Peer Systems (IPTPS'03)*, Berkeley, CA, Feb. 2003.

[162] I. Foster, C. Kesselman, and S. Tuecke. The anatomy of the grid: Enabling scalable virtual organizations. *The International Journal of High Performance Computing Applications*, 15(3):200–222, 2001.

[163] L. Garces-Erice, P. Felber, E. W. Biersack, G. Urvoy-Keller, and K. W. Ross. Data indexing in peer-to-peer DHT networks. In *Proceedings of the 24th IEEE International Conference on Distributed Computing Systems (ICDCS)*, pages 200–208, Tokyo, Japan, 2004.

[164] C. Gkantsidis, M. Mihail, and A. Saberi. Hybrid search schemes for unstructured peer-to-peer networks. In *Proceedings of INFOCOM 2005, 24th Annual Joint Conference of the IEEE Computer and Communications Societies*, volume 3, pages 1526–1537, Miami, FL, March 2005.

[165] M. Harren, J. Hellerstein, R. Huebsch, B. Loo, S. Shenker, and I. Stoica. Complex queries in DHT-based peer-to-peer networks. In *Proceedings of the 1st International Workshop on Peer-to-Peer Systems (IPTPS'02)*, MIT Faculty Club, Cambridge, MA, 2002.

[166] N. J. Harvey, M. B. Jones, S. Saroiu, M. Theimer, and A. Wolman. SkipNet: A scalable overlay network with practical locality properties. In *Proceedings of the 4th USENIX Symposium on Internet Technologies and Systems (USITS'03)*, Mar. 2003.

[167] P. Heleher, B. Bhattacharjee, and B. Silaghi. Are vitrualized overlay networks too much of a good thing? In *Proceedings of the 1st International Workshop on Peer-to-Peer Systems (IPTPS'02)*, MIT Faculty Club, Cambridge, MA, Mar. 2002.

[168] R. Huebsch, J. M. Hellerstein, N. Lanham, B. T. Loo, S. Shenker, and I. Stoica. Querying the Internet with PIER. In *VLDB*, pages 321–332, 2003.

[169] S. Iyer, A. Rowstron, and P. Druschel. Squirrel: A decentralized peer-to-peer web cache. In *Proceedings of 21th ACM Symposium on Principles of Distributed Computing (PODC 2002)*, pages 213–222, Monterey, California, 2002.

[170] X. Jin, W.-P. K. Yiu, and S.-H. G. Chan. Supporting multiple-keyword search in a hybrid structured peer-to-peer network. In *Proceedings of IEEE International Conference on Communications (ICC)*, Istanbul, Turkey, June 2006.

[171] V. Kalogeraki, D. Gunopulos, and D. Yazti-Zeinalipour. A local search mechanism for peer-to-peer networks. In *Proceedings of the eleventh international conference on Information and knowledge management*, pages 300–307. ACM Press, 2002.

[172] D. Karger, E. Lehman, T. Leighton, R. Panigrahy, M. Levine, and D. Lewin. Consistent hashing and random trees: distributed caching protocols for relieving hot spots on the world wide web. In *STOC '97: Proceedings of the twenty-ninth annual ACM symposium on Theory of computing*, pages 654–663, New York, NY, 1997. ACM Press.

[173] J. Kubiatowicz, D. Bindel, Y. Chen, P. Eaton, D. Geels, R. Gummadi, S. Rhea, H. Weatherspoon, W. Weimer, C. Wells, and B. Zhao. OceanStore: An architecture for global-scale persistent storage. In *Proceedings of the Ninth international Conference on Architectural Support for Programming Languages and Operating Systems (ASPLOS 2000)*. ACM, November 2000.

[174] D. Liben-Nowell, H. Balakrishnan, and D. Karger. Observations on the dynamic evolution of peer-to-peer networks. In *Proceedings of the 1st International Workshop on Peer-to-Peer Systems (IPTPS02)*, Cambridge, MA, Mar. 2002.

[175] E. K. Lua, J. Crowcroft, M. Pias, R. Sharma, and S. Lim. A survey and comparison of peer-to-peer overlay network schemes. *Communications Surveys and Tutorials, IEEE*, pages 72–93, 2005.

[176] Q. Lv, P. Cao, E. Cohen, K. Li, and S. Shenker. Search and replication in unstructured peer-to-peer networks. In *Proceedings of the 16th ACM International Conference on Supercomputing (ICS'02)*, New York, NY, 2002.

[177] Q. Lv, P. Cao, E. Cohen, K. Li, and S. Shenker. Search and replication in unstructured peer-to-peer networks. In *Proceedings of the 16th international conference on Supercomputing (ICS'02)*, pages 84–95, New York, NY, 2002. ACM Press.

[178] Q. Lv, S. Ratnasamy, and S. Shenker. Can heterogeneity make Gnutella scalable? In *Proceedings of the First International Workshop on Peer-to-Peer Systems (IPTPS'02)*, pages 94–103, Cambridge, MA, March 2002.

[179] D. S. Milojicic, V. Kalogeraki, R. Lukose, K. Nagaraja, J. Pruyne, B. Richard, S. Rollins, and Z. Xu. Peer-to-peer computing, Mar. 2002.

[180] C. NEWS. Napster among fastest-growing net technologies, Oct. 2000.

[181] C. G. Plaxton, R. Rajaraman, and A. W. Richa. Accessing nearby copies of replicated objects in a distributed environment. In *ACM Symposium on Parallel Algorithms and Architectures*, pages 311–320, 1997.

[182] S. Ratnasamy, P. Francis, M. Handley, R. Karp, and S. Schenker. A scalable content-addressable network. In *SIGCOMM '01: Proceedings of the 2001 conference on Applications, technologies, architectures, and protocols for computer communications*, pages 161–172. ACM Press, October 2001.

[183] M. Ripeanu. Peer-to-peer architecture case study: Gnutella network. In *Proceedings of PDP'02*, Aug. 2001.

[184] M. Ripeanu and I. Foster. Mapping the Gnutella network: Macroscopic properties of large-scale peer-to-peer systems. In F. Kaashoek and A. Rowstron, editors, *Proceedings of the 1st International Workshop on Peer-to-Peer Systems (IPTPS'02)*, March 2002.

[185] R. L. Rivest. The MD5 message-digest algorithm, 1992. Available online at: `http://theory.lcs.mit.edu/~rivest/rfc1321.txt` (Accessed July 24th, 2008).

[186] A. Rowstron and P. Druschel. Pastry: Scalable, decentralized object location, and routing for large-scale peer-to-peer systems. *Lecture Notes in Computer Science*, 2218:329–350, 2001.

[187] A. I. T. Rowstron, A.-M. Kermarrec, M. Castro, and P. Druschel. SCRIBE: The design of a large-scale event notification infrastructure. In *Proceedings of Networked Group Communication*, pages 30–43, London, UK, Nov. 2001.

[188] D. Scholl. Nap protocol specification, 2000. Available online at: `http://opennap.sourceforge.net/napster.txt` (Accessed August 31st, 2007).

[189] R. Schollmeier. A definition of peer-to-peer networking for the classification of peer-to-peer architectures and applications. In *P2P '01: Proceedings of the First International Conference on Peer-to-Peer Computing (P2P'01)*, pages 101–102, Washington, DC, USA, 2001. IEEE Computer Society.

[190] C. Shirky. What is P2P... and what isnt't, 2000. Available online at: `http://www.openp2p.com/pub/a/p2p/2000/11/24/shirky1-whatisp2p.html` (Accessed August 31st, 2007).

[191] M. P. Singh. Peering at peer-to-peer computing. *IEEE Internet Computing*, 1(5):4–5, 2001.

[192] I. Stoica, R. Morris, D. Karger, F. Kaashoek, and H. Balakrishnan. Chord: A scalable peer-to-peer lookup service for internet applications. In *Proceedings of the ACM SIGCOMM'01 Conference*, Aug. 2001.

[193] I. Stoica, R. Morris, D. Liben-Nowell, D. R. Karger, M. F. Kaashoek, F. Dabek, and H. Balakrishnan. Chord: A scalable peer-to-peer lookup protocol for internet applications. *IEEE/ACM Transactions on Networking*, 11(1):17–32, Feb. 2003.

[194] D. Tsoumakos and N. Roussopoulos. A comparison of peer-to-peer search methods. In *Proceedings of the 6th International Workshop on the Web and Databases (WebDB 2003)*, pages 61–66, Mar. 2003.

[195] Z. Xu, M. Mahalingam, and M. Karlsson. Turning heterogeneity into an advantage in overlay routing. In *Proceedings of the Twenty-Second Annual Joint Conference of the IEEE Computer and Communications Societies, INFOCOM 2003*, volume 2, pages 1499–1509, San Francisco, CA, March-April 2003.

[196] B. Yang and H. Garcia-Molina. Improving search in peer-to-peer networks. In *Proceedings of the 22th International Conference on Distributed Computing Systems (ICDCS'02)*, Vienna, Austria, 2002.

[197] B. Y. Zhao, J. D. Kubiatowicz, and A. D. Joseph. Tapestry: An infrastructure for fault-tolerant wide-area location and routing. Technical Report UCB/CSD-01-1141, UC Berkeley, Apr. 2001.

Chapter 5

Grid enabled virtual file systems

5.1 Introduction

The grid is rapidly emerging as the dominant paradigm for wide-area distributed computing [202]. Its goal is to provide an environment for coordinated resource sharing and problem solving in dynamic, multi-institutional virtual organizations [211]. Most of the current grid deployments have focused on data intensive applications where significant processing was done on very large amounts of data [198]. The data required by such applications is largely distributed in various storage systems. The need to access remote data with "near-local" performance is crucial for scheduling and managing of application execution.

One of the grid's purposes is to provide users the ability to share and to use data stored on heterogeneous storage systems as easily as if they were located on a single computer. Unfortunately, this vision is still far from being achieved due to the difficulty to deploy, use and maintain such environments. One of the fundamental problems is the existence of many different administrative domains, different storage systems, different data transfer middleware and protocols in grid environments. This heterogeneity presents an important barrier for data sharing in the grid. Novice grid users, principally scientists who need the power of the grid to solve problems in their own fields, have difficulties in browsing and transferring data. They may find it difficult and cumbersome to write scripts or programs to perform the data transfer between different systems. Data management appears to be a big challenge; it is a time-consuming activity and requires the help of experts with significant expertise in data access related issues.

It is our belief that the widespread uptake of the grid by a wider user community depends on easy-to-use, transparent environments in which users may share and use data collaboratively being unaware of the underlying infrastructure. At the time of this writing, there is a vast volume of projects (e.g., see [209], [199], [220], [212], [203], [230], [207]) that have concentrated their efforts in the development of such environments.

In this chapter, we describe a novel architecture *GRid-enAbled Virtual file sYstem (GRAVY)* [223] [222], which facilitates the collaborative sharing of data in the grid. GRAVY has the following features:

- *Location transparency*: GRAVY allows users to access data that is geographically distributed in multiple domains in the grid without the users having any idea where the data is located.

- *Protocol transparency*: GRAVY provides a generic data transfer architecture that shields users from the complexity of the underlying infrastructures including the system's internal organization and data transfer protocols. As a result, data in heterogeneous file systems can be accessed in a uniform way.

- *Extensibility*: GRAVY allows new protocols to be added as the grid evolves through a set of wrapper interfaces.

The next section of the chapter presents the background for our work including a description of the grid file system, its requirements and an overview of file transfer protocols. Data access problems in grid environments are described in section 5.3, which led to the motivation of our work. Then, we present an overview of related work in section 5.4. Following this, we describe in section 5.5 GRAVY's design and in section 5.6 the architectural issues of the prototype that we have implemented in Java. This prototype allows users to have the view of a unified location-transparent file system of the grid and to access this system without being familiar with the protocol's technical details. Next, in section 5.7 we show the advantages of using GRAVY for data access in a grid environment by two use cases, followed by the experimental results in section 5.8. Finally, section 5.9 concludes the chapter.

5.2   Background

5.2.1   Overview of file system

In computing, the file system is a fundamental data structure that offers data permanent storage. As its name implies, a file system treats different sets of information as directories and files. Each file separates from others in content stored within it and attributes (e.g., file's name, file's access permissions, time and date of the file's creation, access, and modification).

Local file systems are ones that reside entirely on one computer and are accessed from that computer. The local file system provides a namespace, typically represented as a tree. Normally, they form basic blocks of a grid virtual file system.

Grid virtual file systems (GVFS), which store files on one or several heterogeneous storage systems manage resources spread over several autonomous administrative domains. They have to deal with unpredictable performance variations and with changeable system architecture. The GVFS offers a logical resource namespace that allows integrating heterogeneous data storage resources and inter-organizational data.

5.2.2 Requirements for grid virtual file systems

Due to the heterogeneity of the grid, users need a mediating system that serves to abstract the native interfaces of the available resources and provide a common interface. This intermediate system lies below a distributed local file system and is composed of software components that together create an environment where all the resources in the grid are available to the application. This system needs to provide transparent access to data distributed in the grid environment and user interface for users to find files. For example, a user should not have to know where resources are located in order to use them. The system must be able to recover automatically and complete the desired task (i.e., failure transparency). This is the traditional way to mask various aspects of the underlying system.

We present in the following section the set of services that constitutes the fundamental requirements for a GVFS.

Independent multi-institutional data namespace and replication
The dynamic and multi-institutional nature of the grid environment introduces new challenging issues for GVFS in terms of file naming within the grid. GVFS needs a uniform, global, and hierarchical namespace. Uniform and global path names mean that it has transparency of position and location. Transparency of position allows the data consumers (i.e., users and their applications) to be mobile or have multiple positions. Similarly, location independence allows the data file to physically move or migrate from one storage system to others. Providing these capabilities requires an abstraction between the name and location of the data, which allows this mapping to be done dynamically. This separation between the logical name and physical location of the data allows them to be managed separately and there exists no centralized data storage.

Management of global namespaces is fundamental for data location transparency. This global namespace would allow uniform and global path names to persistently address file data within the grid wherever it is located. So that, an application can access data through a logical domain, or global namespace independently with physical data location.

Users can affect the logical view of the data by means of namespace operations such as creation, deletion, and renaming. Infrastructure managers are responsible for deploying and retiring data storage resources, adding and removing data nodes, and configuring networks to optimize a system's ability to handle the demand for data services.

Availability of a requested data item is an important performance parameter. A well-known technique for improving availability in distributed systems is replication. If multiple copies of data exist on independent nodes, then the chances of at least one copy being accessible are increased. Aggregate data access performance will also tend to increase, and total network load will tend to decrease, if replicas and requests are reasonably distributed.

Secure access GVFS should provide access to distributed files hosted on heterogeneous storage systems with different security domains and support sharing of data for processing and large-scale collaboration. Moreover, it must allow users and applications to gain secure access to remote data resources. For example, two collaborators at sites A and B need to share the results of a computation performed at site A, or perhaps design data for a new part needs to be accessible by multiple users at different sites. In this case, it's necessary to have a mechanism to authenticate and identify users (i.e., the processes acting on their behalf). Concretely, GVFS must address local security integration, secure identity mapping, secure access/authentication, secure federation, and trust management.

GVFS must support global authentication (i.e., single sign-on) with identities that span administrative domains and organizations, support the establishment of virtual organizations (i.e., groups that span organizations), enforce access control policies, and protect data on storage systems. And it should guarantee that the sites retain full control over their resources. The data transfers must be protected using secure communication channels.

Interfacing to mass storage management systems The ability to access data at a wide variety of heterogeneous storage systems, which are different in architecture and administration policy, in a standard way will facilitate inter-operability in data management and access. A uniform access interface allows a uniform and coordinated access to multiple data resources. The underlying file system resources could have their own transport, access, authentication/authorization protocols that lead to proprietary protocol clients. As a result, a common access interface is needed for decoupling users from data access protocols.

5.2.3 Overview of file transfer protocols

HyperText Transfer Protocol - HTTP The HTTP protocol has rapidly become one of the major protocols used for inter-computer communications on the Internet. It is designed with the lightness and speed necessary for distributed, collaborative, hypermedia information systems. HTTP is a generic, stateless, object-oriented protocol which can be used for many tasks, such as name servers and distributed object management systems, through extension of its request methods (commands). With HTTP protocol, a client makes a TCP connection to a server, sends it a simple-to-parse string containing a command along with a few parameters, and receives the response over the same connection.

File Transfer Protocol - FTP The classic File Transfer Protocol (FTP) is the most commonly used protocol for data transfer on the Internet. It defines a capability for machines to send and receive data using the underlying TCP/IP protocol of the public Internet. FTP, which is an open standard and widely deployed in almost every operating system, is well documented and broadly accepted as the de-facto data movement mechanism in large-scale networks.

The FTP protocol is inherently vulnerable to snooping attacks since passwords are transmitted in the clear, as is data. Since FTP requires making a copy of the data at a remote machine, if the original file is ever changed, the new version of the file needs to be updated on the remote machine. This process is fraught with the potential for inconsistencies. Also, FTP is an all-or-nothing protocol - if even one bit of a large file changes, the entire file must be copied over. Finally, FTP is not conducive to programmatic access. Therefore, applications cannot take advantage of remote files using FTP without significant change.

GridFTP GridFTP extends the existing FTP protocol with new features required for large volume, fast data transfer in the grid with the grid security model. GridFTP supports parallel data transfer by using multiple parallel TCP streams, which can improve aggregate bandwidth over using a single TCP stream. However, this functionality limited the available transfer bandwidth to the maximum network bandwidth between the sender and the receiver of the transfer in the grid. Moreover, many applications require transfer of only a portion of a file because transferring the entire file could be too expensive.

GridFTP offers a useful feature called *striping* to enable the high-performance data transfer using multiple hosts to move a single file. With striping, it is possible for multiple nodes to work together as a single GridFTP server. Each node can read and write only the pieces of the file that it is responsible for. In that way, transfers are divided over all available back end data nodes, thus allowing the combined bandwidth of all data nodes to be used.

In order to manage large data sets it is necessary to provide third-party (direct server-to-server) control of transfers between storage servers. GridFTP provides this capability by adding GSSAPI security to the existing third-party transfer capability defined in the FTP standard.

GridFTP solves the privacy and integrity problems with FTP by encrypting passwords and data. Moreover, GridFTP provides for high-performance, concurrent accesses by design. An API enables accessing files programmatically, although applications must be re-written to use new calls. Data can be accessed in a variety of ways, for example blocked and striped. Part or all of a data file may be accessed, thus removing the all-or-nothing disadvantage with FTP.

SSH/SCP SSH, the Secure Shell, is a very common protocol to create a channel for running a shell on a remote computer, with end-to-end encryption between the two systems. As a result, data is automatically encrypted whenever it is sent over the network. When the data reaches its recipient, it is automatically decrypted. This transparent encryption enables users to work normally, unaware that their communications are safely encrypted on the network.

SCP, Secure Copy, which belongs to the SSH protocol, allows files to be transferred securely between remote sites. For example, a user wishes to transfer a file across two remote sites. The file contains business secret information that must be kept from prying eyes. A traditional file-transfer program, such as FTP, doesn't provide a secure solution. A third party can intercept and read the packets as they travel over the network. Using SSH, the file can be transferred securely between machines with a single secure copy command. When transmitted by SCP, the file is automatically encrypted as it leaves the sender site and decrypted as it arrives on the receiver site.

5.3 Data access problems in the grid

A grid is a heterogeneous environment

A frequent obstacle for data-intensive applications to operate effectively in grid environments is access to remote data. This problem is challenging because the grid is a heterogeneous environment. Data at each site is accessed through different mechanisms including how the data is organized, which transfer protocols are supported, and how the authentication is carried out. Users are forced to deal with such aspects whenever they want to access data at different storage systems and it is difficult to efficiently share data between these systems.

We considered that data resides in different forms, ranging from structured data organized in relational database systems to unstructured data organized in file systems. Some works are currently ongoing to integrate structured data on the grid [199], [231]. Here we propose the GRAVY architecture that allows the inter-operability of unstructured data on the grid. In other words, distributed file systems will be inter-operable irrespective of their heterogeneity.

Grid jobs need distributed data to run

In order to run grid data-intensive jobs, the input data need to be transferred to the appropriate location at the time the computation needs it. This task is commonly referred to as file *stage-ins*. The output data is moved back to its home storage systems as the computation is completed. This task is commonly referred to as file *stage-outs*.

In the grid, and on the Internet, files are accessible through a variety of different protocols supported by storage systems, such as HTTP [206], FTP [225], SCP/SSH [233], and GridFTP [198]; each has its own data interaction styles. For example, a lot of tools supporting the protocol FTP provide an easy-to-use graphical interface such as SmartFTP [227], FileZilla [208], SecureFTP [226]. Usually, due to security problems, the protocol SCP/SSH is preferred instead of FTP. SCP is the protocol that allows files to be transmitted with the encryption benefits of protocol SSH. Most SSH client tools include SCP capability through a command-line utility. GridFTP provides secure and efficient data movement in the grid environment. This protocol, which extends FTP, is developed as part of the Globus Project [216]. Unfortunately, there are few tools available supporting this protocol. In current practice, the popular choice for data access in GridFTP is using APIs supported by GT4 [209].

The diversity of data interaction styles (e.g., GUI, command-line, APIs) forces users to switch from one interaction style to another for file staging between heterogeneous systems. Hence, it prevents the automation of data transfers. Some interaction styles, such as GUI and command-line, are only intended for manual use or simple scripts. Others, such as APIs or Web services, allow file staging to be performed in programs. Due to this diversity, users are obliged to manually transfer files between heterogeneous systems by using different tools or writing scripts and programs to perform file staging. Manual file staging is not suitable for applications in grid environments as it supposes users to know in advance which files will be needed during the computation. Generally, users don't have the knowledge of the server that will be chosen for the computations. The choice of computational server is done by the job scheduler. So, it is important for job scheduler to have a mediating system that is able to control the placement of data needed for the computation. The automation of data transfers becomes a crucial factor for data management system in the grid.

5.4 Related work

A number of initiatives to address data management in grid environments have been initiated in recent years. We describe below some of these initiatives.

Based on the basic Globus services [216], the DataGrid [203] is a large and complex project that defines a layered architecture of service components for transferring large datasets in heterogeneous environments. This architecture is similar to ours (i.e. GRAVY) in the sense that both try to separate the physical location of data from its logical view, which is called metadata.

GT4 [209] provides a number of components for data management. These components fall into two basic categories: data movement, which is composed of GridFTP tools and Reliable File Transfer (RFT) service, and data replication, which consists of Replica Location Service (RLS). An important related component, OGSA-DAI [199], provides data access and integration capabilities to data resources, such as databases, within a WebService-based framework.

LegionFS [230] proposes a virtual file system based on NFS protocol. The core of LegionFS functionality is based on an object-based system that employs a basic object providing access methods similar to UNIX system calls (e.g., read, write, seek). NFS requests from the client will be interpreted to appropriate methods of this basic object.

Within the EGEE project [205], the data management system (DMS) [220] is composed of several components. The first is storage elements (SEs) which are the real element doing the storage of files. In the framework of the DMS, files are available through two namespaces: logical (Logical File Name - LFN) and physical (Storage File Name - SFN). The DMS is responsible for mapping an LFN to one or more SFNs. Other components of DMS are data catalogs that offer access to file replicas using LFN and a data scheduler, which assures the availability of data at the chosen site for computation.

A standardization effort of the Global Grid Forum Grid File System working group (GFS-WG) [213] is to provide a service-oriented architecture for a Grid File System (GFS) [214] that provides standard interfaces to facilitate the federation and sharing of virtualized data. It should be noted that GFS is a specification, not an implementation.

Adapting Peer-to-Peer data transfer methods, [228] and [229] propose to use BitTorrent as a protocol for large file transfers in the context of desktop grids. It is shown that BitTorrent is efficient, scalable when the number of nodes increase, but suffers from a high overhead when transmitting small files. The papers investigate the approaches to overcome these limitations.

Compared with GRAVY, these solutions are designed to work primarily with their own self-contained middleware, (e.g., LegionFS in Legion middleware, DMS in gLite, RFT in GT4) or suppose to use a principal protocol for data transfers in the grid (e.g., BitTorrent). On the other hand, GRAVY is designed to integrate into any global scheduling systems and an important feature of GRAVY is that it supports multiple protocols at both the server side and remote side.

5.5 GRAVY: GRid-enAbled Virtual file sYstem

5.5.1 Design overview

As indicated previously, the heterogeneity of storage systems presents an important brake on the collaborative use of data by a wider community in the grid. Currently, data access is restricted to expert communities who have a deep knowledge of file organization and data access protocols on individual storage systems. However, the primary users of grids are domain scientists and engineers who have little expertise in coping with these issues. These users need to access data in an easy way despite the distributed and heterogeneous nature of storage systems. They should not be forced to know a storage system's internal organization or the transfer protocols' technical details.

In order to mask the heterogeneity of storage systems, our approach is to build a virtual file system GRAVY on top of underlying file systems. This virtual file system allows data to be transferred on-demand between heterogeneous file systems in a uniform fashion irrespective of its access protocol. Figure 5.1 shows the conceptual overview of GRAVY. In the next section, we describe these components separately and how they work together through an example of user interaction.

We refer to users who are expert on data access protocols as *file-system-providers*. Their role is to make their file system accessible to users through GRAVY. GRAVY appears as a general data broker that negotiates with the underlying file systems the availability and exploitation of data. Concretely, all interactions of GRAVY with the underlying file systems are mediated through wrapper interfaces.

5.5.2 Component description

The core services of GRAVY are composed of a virtual layer and core layer which consist of four major components: *virtual interfaces*, *TransferManager*, *AccessManager* and *wrapper interfaces*. Their role is to provide the user layer with uniform and seamless access and management of data transfers between remote file systems on the physical layer.

The *virtual interfaces*, which consist of *GridFileSystem* and *GridFile* are designed to simplify and unify the way in which users handle data from heterogeneous data sources. The user layer is able to remotely interact with the virtual interfaces through a variety of supported access protocols, including HTTP, FTP, and Web services. Local access to virtual interfaces is possible through a set of APIs that allow applications or job schedulers to control data placement.

The core layer is composed of four components: the *FileActionQueue*, the *TransferManager*, the *AccessManager* and the *wrapper interfaces*. User re-

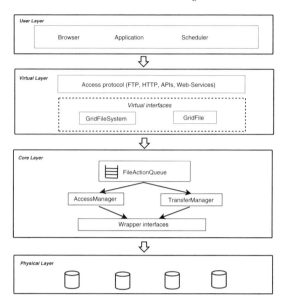

FIGURE 5.1: Conceptual design of GRAVY. The dashed rectangle is the core services of GRAVY.

quests received from the virtual interface are queued in *FileActionQueue*, which examines each request in order to route each correctly to the *Transfer-Manager* or the *AccessManager*.

We classify the user requests in two categories: transfer requests and access requests. Transfer requests need to be treated differently from access requests, since transfer requests generally have long execution time and they can fail for a variety of reasons at anytime during the execution. They need to be monitored and rescheduled for restart in case of failure. Hence, the *Transfer-Manager* is designed to execute transfer requests asynchronously. The *TransferManager* performs the movement of files from one remote file system to the other. In case of transfer failure due to dropped connections, machine reboots or temporary network outages, the *TransferManager* will restart the transfers at another time in order to assure the successful completion of transfers. In contrast, the access requests (e.g., directory creation, file rename) have a short execution time, so the *AccessManager* is designed to execute access requests synchronously. It performs access operations on the remote file systems and returns immediately to users the result of execution.

The *TransferManager* and the *AccessManager* interact with the remote file systems through *wrapper interfaces*. These interfaces are implemented by the file-system-provider in an appropriate protocol that is specific for each file system.

5.5.3 An example of user interaction

We now describe how these components work together through an example interaction scenario. We consider a sequence of two user requests: a directory creation (i.e., an access request) and then a file copy (i.e., a transfer request) into this directory. Firstly, the user connects to GRAVY with a request to create a new directory. This request in a specific access protocol (e.g., *MKD* in FTP) will be translated into a common access request (e.g., *mkdir* method) for the virtual interface *GridFile*. This request, after being queued in the *File-ActionQueue*, is routed to the *AccessManager*. The *AccessManager* demands the *wrapper interfaces* to create a connection with the appropriate remote file system. After authenticating to the remote file system, the *AccessManager* creates the new directory synchronously and sends the result of execution back to the user through the virtual interface.

After knowing that the directory is created successfully, the user sends a request to transfer a file into this directory. Similarly, this transfer request (e.g., *STOR* in FTP) is translated into a common transfer request (e.g., *copyTo* method) for the virtual interface *GridFile* and then queued in *FileActionQueue* before being sent to the *TransferManager*. Firstly, the *TransferManager* connects to the remote file system through the *wrapper interfaces* and asks for the transfer permission. Then, it performs the transfer asynchronously and informs the transfer status to the user if needed. In case of transfer failure, the *TransferManager* assumes the responsibility of restarting the failed transfer to assure that the transfer is successfully completed.

5.6 Architectural issues

5.6.1 Protocol resolution

GRAVY supports multiple access protocols on both the server side and remote side (see Figure 5.2). This is a crucial requirement of a virtual file system used in a heterogeneous grid environment. Although GridFTP is promoted as a basic grid protocol for transferring data between grid nodes, there exist a large number of existing file systems supporting other protocols. For example, wide-area file access is still likely to be dominated by FTP, SSH protocols.

Server side

At the server side, supporting multiple protocols not only allows users to use their preferred file transfer protocol to interact with GRAVY but also allows GRAVY to be easily and flexibly deployed according to user needs. For example, HTTP access allows GRAVY to integrate easily into web portals of the grid. Local access via APIs and Web services access allow GRAVY to

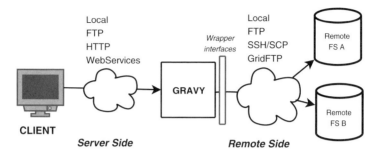

FIGURE 5.2: Multiple access protocol in both server side and remote side.

integrate into applications and the job scheduler for data placement control. Besides local access, GRAVY currently supports three protocols: FTP [225], HTTP [206], and Web services. The implementation of FTP access is based on [201]. The Web services protocol is deployed using WSRF framework implemented in GT4 [209].

Remote side

At the remote side, supporting a variety of access protocols allows GRAVY to support a large number of existing file systems. Although GridFTP has been promoted as the standard protocol for data movement in the grid, there is a large number of existing file systems supporting other protocols.

From the file-system-provider's point of view, the remote file system is simply a storage system abstracted into directories and files and supported by an access protocol (e.g., FTP, GridFTP, HTTP) or a file server in other words. In order to make a file system inter-operable with others, the user needs to develop connectors between protocols supported by this system to all existent protocols in the grid (see Figure 5.3a). This practice is not suitable for the continually evolving grid architecture as it requires adding a new protocol connector if a file system-support new protocol is integrated to the grid. In GRAVY, this task is simplified by the *wrapper interfaces* that are in charge of creation and management of connections between GRAVY and remote file systems. *Wrapper interfaces* play the role of a bridge between GRAVY and remote file systems. They make GRAVY completely modular; it is easy to add support to GRAVY for a new protocol (see Figure 5.3b).

The wrapper interfaces are composed of the four interfaces: RemoteFile, RemoteFileSystemDriver, RemoteFileSystemConnection and FileTransfer. Figure 5.4 shows the structure of wrapper interfaces with a list of methods for each interface.

- *RemoteFile*: as its name implies, contains methods for retrieving information about the remote file object on the remote file system (e.g., name, size, latest modified time, control permission).

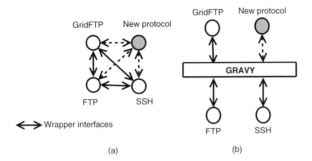

FIGURE 5.3: Integration of new protocol at the remote side in GRAVY.

- *RemoteFileSystemDriver*: contains access information of the remote file system (e.g., supported protocol name, protocol port number, file system's address) and returns a *RemoteFileSystemConnection*'s instance for the virtual layer to interact with the remote file system.

- *RemoteFileSystemConnection*: contains control methods (e.g., open, close, ping) of GRAVY's connection to the remote file system and implements methods of access operations to files that reside on that remote file system (e.g., creating a new directory, listing the files present in a directory, deleting files and directories).

- *FileTransfer*: contains file transfer implementation in a specific protocol. Two types of connections specified are needed for transfer operations: two-party transfer represented as DefaultFileTransfer class and third-party transfer represented as ThirdPartyFileTransfer class. These classes must be threads; they must implement the Runnable interface because users will not deal with details in the time dimension with the file after they have decided to transfer a file. Since DefaultFileTransfer and ThirdPartyFileTransfer have many common properties structurally and behaviorally, they are extended from the FileTransfer class and override necessary methods.

Besides the implementation of wrapper interfaces for local file systems, we have used client-side libraries provided in GT4 [209] to implement *wrapper interfaces* for FTP and GridFTP protocol, and JSch[218] for SSH protocol. It is the role of the file-system-provider to implement the wrapper interfaces in order to integrate a new protocol in GRAVY.

Security

Security is an important issue on the grid due to different administrative domains and policies. Since each protocol has its own authentication mech-

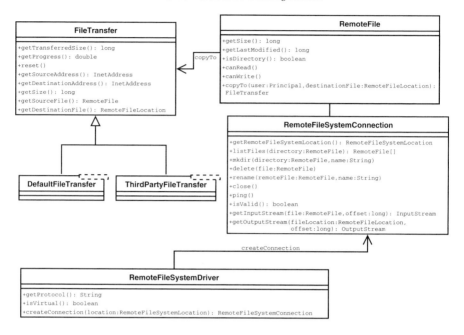

FIGURE 5.4: Class diagram of the wrapper interfaces.

anism, it enforces its own access control policy. This results in difficulty in establishing confidence across different protocols. Our solution is to adopt the Grid Security Infrastructure (GSI) provided by Globus [210] because it avoids a centrally-managed security system and supports *single sign-on* for users of the grid. For other protocols, authentication is performed through anonymous access.

GSI is based on public key encryption, X.509 certificates, and the Secure Sockets Layer (SSL) communication protocol. Each grid user is provided a unique identity within the virtual organization that he/she belongs to. Access control of local resources is done by the mapping between the user's unique identity and local user identity in a global configuration file called *gridmap-file*. We have implemented security by wrapping GRAVY calls with GSI. GSI maps a user global identity to a local user and GRAVY checks the permissions of files to see whether a user has enough privileges to access the file.

5.6.2   Naming management

In a grid environment, management of data across multiple virtual organizations presents challenging problems for data naming. The Resource Namespace Service (RNS), a specification of the Grid File System working group of the Global Grid Forum [224], is proposed to provide a naming mechanism to link existing data sources. RNS proposes a three-tier naming architecture

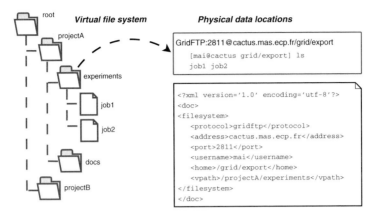

FIGURE 5.5: Example of a logical view and its mapping to physical data locations.

that consists of human interface names, logical reference names, and end-point references. Mapping from a human readable name to an actual data location can be realized in two levels of indirection. The first level is mapping human interface names directly to end-point references. The second level is realized by mapping human interface names to logical reference names (that may not be very readable by humans), which in turn map to end-point references.

In GRAVY, we applied the first level of indirection for the naming management. The *GridFileSystem* interface is responsible for decoupling the logical view of the data from its physical location. This interface represents the virtual global file system with hierarchical organization of virtual directories where leaves in this tree correspond to physical data locations on a remote file system. Users can create their own logical view of grid data where a logical directory may not necessarily correspond to the physical directory. Different users have a different logical view if they have different rights on data resources.

The *GridFileSystem* instance is specified using a configuration file written in XML and is initialized at runtime. Figure 5.5 shows an example of a logical view of grid data and its mapping to physical data locations with an example configuration file. In this example, the *export* directory on server *cactus* is mounted to the *experiments* directory of the virtual file system. The files under the *experiments* directory (e.g., *job1* and *job2*) are assumed to be under the corresponding physical directory. The resolution of *experiments* directory's content is performed only at the runtime.

5.6.3   GridFile - virtual file interface

The fundamental requirement for virtual file systems used in the grid is that all these file operations in different protocols must be made completely transparent to users. Accessing the local file system for listing files, changing

```
1: GridFile root = GridFileSystem.getRoot();
2: RemoteFileSystemLocation remoteFileSystem =
3:     GridFtpRemoteFileSystemDriver.getLocation(InetAddress.getByName("cactus.mas.ecp.fr",
                                     2811,"mai","/grid/export");
4: GridFile theDir = root.mkdirs("/projectA/experiments");
5: theDir.mount(remoteFileSystem);
6: theDir.listFiles();
7: GridFile destination = root.resolve("/projectB");
8: theDir.copyTo(destination);
```

FIGURE 5.6: Example of using GridFile's methods.

directories, etc. should be no different than accessing any remote file system with any access protocol. Transfer operations (e.g., *copy*, *move*) must be as applicable to local files as they are to data hosted on remote file systems. With these concerns in mind, we design the *GridFile* as a virtual file object that provides the single consistent protocol-independent interface for access and transfer operations in the virtual file system. This uniform interface, which provides a set of generic file operations, should keep the user shielded from protocol peculiarities. GridFile contains also the basic properties such as size, name, modification date, access permissions, etc. of the remote file. A non-exhaustive list of GridFile's operations is shown in Table 5.1. These generic operations correspond easily to popular file operations in different protocols since most file operations across protocols are very similar (e.g., all have directory operations such as *create*, *delete* as well as file operations such as *read*, *write*). Administrative methods allow users to interact on the logical structure of the virtual file system. Access methods and transfer methods, as their names indicate, allow users to send access or transfer requests to remote file systems. Figure 5.6 presents an example of using GridFile's methods.

As can be seen from Table 5.1, some operations such as *copyTo()*, *moveTo()* and *delete()* can be managed asynchronously because they may take a long time to complete. Their goal is to prevent users from being locked while waiting for the operations to finish.

5.6.4   Data access

The AccessManager is responsible for carrying out the access requests and returns the result to the virtual layer. The AccessManager translates these requests into the specific protocol supported by the remote file system and accomplishes it by interacting through the *wrapper interfaces*. Figure 5.7 shows how the AccessManager handles an access request of the GridFile interface.

1. Suppose that the user calls the *listFiles()* operation of the GridFile interface to get a list of files of this virtual directory.

2. The GridFile adds the request to the request queue of the AccessManager.

Table 5.1: Supported methods of GridFile interface.

Administrative-related methods	
mount	Mounts a logical name with a physical data location
resolve	Resolves a grid path to a grid file object
refresh	Refreshes current logical view
Access-related methods	
listDirectories & listFiles	Retrieves and lists all the subdirectories and files in human readable format
delete & asynchDelete	Deletes a file or a directory. This operation can be performed in asynchronous mode
mkdir(s)	Creates a directory or directories
renameTo	Renames a directory or a file
getName	Returns name of current directory or file
getLastModified	Returns the last modified time of current directory or file
getSize	Returns size of current directory or file
Transfer-related methods	
copyTo & asyncCopyTo	Transfers directories or files to another specified grid destination. When performed in asynchronous mode, it returns an instance of *GridFileTransfer* containing control methods (e.g., *start*, *cancel*, *pause*, *getProgress*, *getStatus*) for a file transfer
moveTo & asyncMoveTo	Moves directories or files to another specified grid destination. When performed in asynchronous mode, it returns an instance of *GridFileTransfer*
getInputStream	Gets input stream of a file
getOutputStream	Gets output stream of a file

3. The AccessManager invokes the *getConnection()* operation on Remote-FileSystemDriver to get a connection between GRAVY and the remote file system.

4. The AccessManager receives the RemoteFileSystemConnection object.

5. The AccessManager performs the *listFiles()* operation on Remote-FileSystemConnection in a specific protocol.

6. Once the *listFiles()* operation finishes, the result is sent back to the AccessManager.

7. The AccessManager forwards the result to the GridFile interface.

8. The GridFile informs the user of the execution results or sends appropriate error messages if there are errors.

9. Once the request execution finishes, the AccessManager releases the connection.

10. Finally, the RemoteFileSystemConnection is closed.

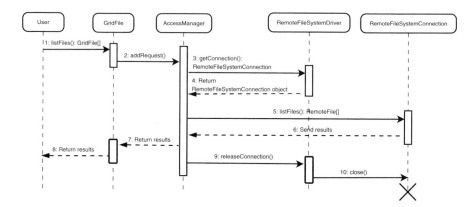

FIGURE 5.7: The sequence diagram for AccessManager.

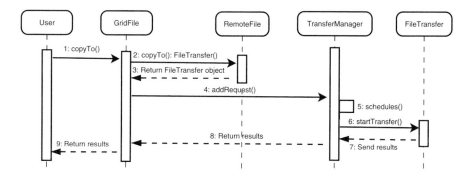

FIGURE 5.8: The sequence diagram for the execution of a transfer request in synchronous mode.

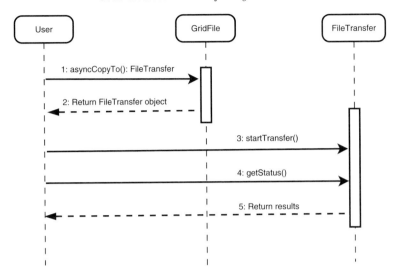

FIGURE 5.9: The sequence diagram for the execution of a transfer request in asynchronous mode.

5.6.5 Data transfer

The TransferManager takes care of transferring files between remote file systems. It manages file transfers in different protocol connections to allow transparent two- and three-party transfers. In order to support multiple concurrent file transfers, each transfer is launched as a thread, which can be performed in synchronous or asynchronous mode. Figure 5.8 illustrates the sequence diagram for the execution of a transfer request in synchronous mode (e.g, *copyTo()*).

1. Suppose that the user calls the *copyTo()* operation of GridFile interface to perform a file transfer to another virtual file reference.

2. As a GridFile points to a RemoteFile which is a physical file on the remote file system, the GridFile invokes the *copyTo()* operation of RemoteFile. This returns a FileTransfer object that contains the information required for performing file transfers (e.g., protocol name, source and destination address, file name).

3. The GridFile receives the FileTransfer object.

4. The GridFile adds the FileTransfer object to the request queue of the TransferManager.

5. The TransferManager schedules the transfer requests, for now it uses a *"the first-come, first-served"* strategy to execute these requests.

6. Then, the TransferManager performs the transfer by invoking the *start-Transfer()* method of FileTransfer. Then, this object initiates a third-party transfer using ThirdPartyFileTransfer class or opens two connections, one from the source and one to the destination file system for file transfers using DefaultFileTransfer class.

7. The result of file transfer is sent back to the TransferManager.

8. The result is then forwarded to the GridFile interface.

9. Finally, the GridFile informs the user of the execution results or sends appropriate error messages if there are errors.

The Figure 5.9 presents the sequence diagram for the execution of a transfer request in asynchronous mode (e.g, *asyncCopyTo()*). In this case, it is not the role of TransferManager but of the user to control the transfer process.

1. The user calls the *asyncCopyTo()* operation of GridFile interface to get a FileTransfer object, which implements the Runable interface to be a thread.

2. The user receives the FileTransfer object.

3. The user invokes the file transfer on FileTransfer object. GRAVY launches a new thread to execute the file transfer, so the user is not locked waiting for completion.

4. In order to know the status of the file transfer, the user can call the *getStatus()* method to get the current status of the transfer.

5.7 Use cases

5.7.1 Interaction with heterogeneous resources

We consider a simple use case when a user wishes to access files in two different file servers (*cactus* and *tulip*) as shown in right part of Figure 5.5. Firstly, he/she can choose his/her preferred protocol to connect to GRAVY. In this use case, we suppose he/she uses local access to communicate with GRAVY. After he/she obtains a reference of root *GridFile* object: *refRoot*, he/she can perform a *refRoot.listFile()* operation to get the list of directories: *projectA* and *projectB*. It should be noticed that these directories are completely virtual; there are no physical locations corresponding to these directories. *projectA.listFile()* will list child virtual directories: *experiments* and *docs*. Since the *experiments* directory is mapped to a physical data location on the *cactus* server, the *experiments.listFile()* operation will ask the *Access-Manager* to connect to the *cactus* server in GridFTP protocol and to convert the virtual path (*/projectA/experiments*) to a physical path (*/grid/export*) to obtain the list of children of that path: *job1* and *job2*. In this simple use case, the advantage is the simplicity: users can use any protocol client to communicate with the virtual grid file system, and data interaction on underlying heterogeneous file systems (e.g., *cactus* and *tulip*) is completely transparent to users.

5.7.2 Handling file transfers for grid jobs

We will show in this use case the contributions of GRAVY in handling file transfers for grid jobs. We suppose a case when a user wants to submit a staging job that uses files on his/her computer.

Without-GRAVY scenario: the user has to use a job description language (e.g., RSL [215] for Globus) to specify information related to job submission (e.g., jobType, fileStageIn, fileStageOut). Then, he/she sends this RSL file to the compute node. All the job arguments contained in the RSL file (stage-in files, parameters, etc.) will be transferred to the compute node in advance of the job execution. The files that are created or modified by the job will be returned to the user if they are specified as stage-out files in RSL file. In this scenario, the user has to perform discovery and choice of compute node in order to transfer to it the stage-in files that are necessary for job execution.

With-GRAVY scenario: supposing that GRAVY is integrated into a high-level scheduler, the user needs to specify only the input files for the job execution, which are located on his/her computer. It is the role of the high-level scheduler to discover and choose the compute node for the user's job. Thanks to GRAVY, the scheduler can perform file transfers from the user machine to the chosen compute node irrespective of protocols they support. Since a job can have its own view of the file system, file-path references of stage-in files in the user local namespace (e.g., */grid/export/job1*) can be passed to the job as arguments. GRAVY will create the same virtual file-path and map a physical file location to this file-path. When a job performs a file action (e.g., read, write, copy) to this virtual file-path, GRAVY will perform corresponding file actions on physical files.

This use case shows the advantages of using GRAVY in handling file transfers for grid jobs. Firstly, a job with staging files can be separated into transfer jobs and computational jobs. In that way, transfer jobs can be queued, scheduled, monitored and managed asynchronously from computational jobs. Secondly, GRAVY allows the user's file system view to be the same regardless of the file locations. This feature enables transparent grid file access for any middleware (i.e., applications, schedulers) that uses GRAVY.

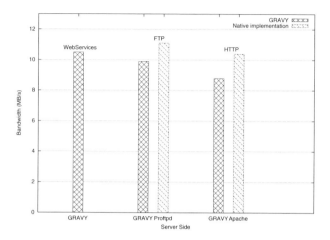

FIGURE 5.10: Server side results.

5.8 Experimental results

GRAVY's latest version runs on any platform that supports the Java VM 5.0. Firstly, we perform a series of data transfers to test GRAVY's feature of supporting multiple protocols. Secondly, in order to evaluate the processing efficiency and performance of our prototype, we perform a set of concurrent file transfers and use the modified Andrew benchmark [217] that is the well-known benchmark to test the performance of a distributed file system. The benchmark consists of five phases: (i) create directories, (ii) copy files into the directories, (iii) list file attributes, (iv) scan the files and (v) compile the files.

The experiments are performed on four Pentium 4 3.2 GHz machines with 512 MB of RAM, each running Linux with kernel 2.4.x. They are directly connected to 100 Mbps network adapter.

5.8.1 Support for multiple protocols

We perform file transfers at both the server side and remote side in different protocols. The experimental setup is shown in Figure 5.2. "Server side" means that the transfers occurred between the client and GRAVY. "Remote side" means that the transfers are launched by GRAVY to move data between remote file systems. At the server side, we compare the bandwidth delivered to the client by GRAVY to that delivered by native implementation of each protocol. At the remote side, we observe the bandwidth obtained for each change of protocol at remote file systems.

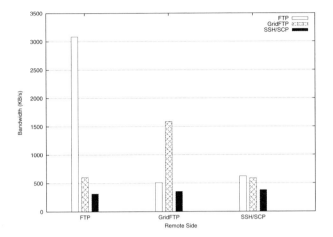

FIGURE 5.11: Remote side results.

Server side

In the first sets of experiments, our goal is to illustrate that the bandwidth delivered by GRAVY at the server side is very similar to that of the native server. The client asks GRAVY to transfer a file of 50MB in FTP, HTTP and Web services respectively. Then we repeat the above transfer using a native protocol server (i.e., ProFTP for FTP and Apache for HTTP) to evaluate the bandwidth delivered by GRAVY. The results in Figure 5.10 show that the bandwidth delivered by GRAVY is just a little lower than the one of the native servers.

Remote side

We perform file transfers of of 10MBs from file server A to file server B (see Figure 5.2) using different protocols. The transfers in GridFTP and FTP are repeated with *globus-url-copy* command-line utility supplied with Globus Toolkit to compare with the bandwidth delivered by GRAVY. The results in Figure 5.11 are the average of 10 file transfers. We observe that the bandwidth varies a lot across each change of protocol at the remote file system. We get better bandwidths for the transfers using the same protocol. The only exception is the transfers in SSH protocol; the reason is that this protocol doesn't support third-party transfers like FTP or GridFTP. We note that the bandwidth of GRAVY for transfers in FTP and GridFTP is very similar to that of the *globus-url-copy* tool.

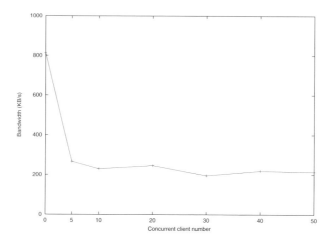

FIGURE 5.12: Processing performance of GRAVY depending on the number of clients concurrently transferring files.

5.8.2 Performance

5.8.2.1 Many concurrent file transfers

In order to test the stability and processing efficiency of GRAVY, we write a client program using GRAVY to launch several concurrent processes reading a remote file into a buffer and writing the data out to a local file. The tests were done with files of 10MB. The result as the transferred KB per second depending on the number of concurrently connecting clients is shown in Figure 5.12. Each value is an average of 5 tests. It shows that GRAVY has a problem with many concurrent requests. It is predictable that GRAVY achieves high performance for low numbers of connecting clients. For increasing number of concurrent clients, its performance decreases smoothly but it remains relatively stable.

5.8.2.2 Andrew benchmark results

We use the modified Andrew benchmark to compare GRAVY's performance to the Linux 2.4.x local file system and NFS v3. For the NFS measurements, we run the benchmark on a NFS client accessing a single NFS server. For the GRAVY measurements, we implemented a Java program that performs a pattern of file system accesses equivalent to the one of the Andrew benchmark because the current prototype implementation of GRAVY provides only Java interfaces to the file system. We repeat the execution of our Andrew-like Java program on GRAVY with three different configurations. Concretely, the directory on which we run the benchmark is mounted to a different remote file system for each execution. The remote file system is accessible in GridFTP, FTP and SSH protocol respectively. Files used during the compilation phase

Table 5.2: The Andrew benchmark results on Linux 2.4.x local file system, NFS and GRAVY. Each table entry is average elapsed time in milliseconds of five runs of the benchmark. The rightmost column shows the average elapsed time of the benchmark runs on GRAVY with three different configurations.

| Phase | Local | NFS | GRAVY | | |
			GridFTP	FTP	SSH	Av-erage
1	8.04	361.66	96128.00	4172.20	17277.60	39192.60
2	93.32	3293.31	194150.60	18635.40	100861.80	104549.27
3	237.48	2856.21	50848.00	4397.00	39008.20	31417.73
4	298.43	3466.46	17142428.80	15837.40	165160.80	117475.67
5	3773.75	4552.05	4015.20	3985.60	4038.60	4013.13
Total	4411.03	14529.68	516570.60	47027.60	326347.00	296648.40

are stored locally for remote access on these remote file systems. The directory that we use as input to the benchmark contains 15 directories and 96 C source and header files for a total size of 511KBs. Table 5.2 shows the results of running the Andrew benchmark on the Linux 2.4.x local file system, NFS and GRAVY.

As expected, the local file system has the best performance on all five phases because it performs no network communication. The benchmark results on GRAVY have a high variance for each configuration. We achieve better performance with FTP configuration, followed by SSH and GridFTP configurations respectively. In the compilation phase, all file systems achieve a very similar performance because the performance of this phase is primarily limited by the speed of the CPU. For the other phases, GRAVY is slower than NFS due to the time needed for the authentication and the resolution between logical names and physical data locations.

5.9 Concluding remarks

In this chapter, we have introduced GRAVY, a grid-enabled virtual file system, which enables the inter-operability between heterogeneous file systems in the grid. We have pointed out the current challenges for data access in the grid and how GRAVY can provide solutions to them. GRAVY integrates underlying heterogeneous file systems into a unified location-transparent file system of the grid. This virtual file system provides applications and users a uniform global view and a uniform access through standard APIs and interfaces.

With two use cases, we have shown the contributions of GRAVY in solving data access problems in a grid environment. Our approach is validated by a prototype implemented in Java. This prototype shows that the way users access data is simplified and that data transfers between heterogeneous file systems can be automated. This feature allows GRAVY to integrate with a high-level scheduler for handling data transfer jobs.

References

[198] B. Allcock, J. Bester, J. Bresnahan, A. L. Chervenak, I. Foster, C. Kesselman, S. Meder, V. Nefedova, D. Quesnel, and S. Tuecke. Data Management and Transfer in High Performance Computational Grid Environments. *Parallel Computing Journal*, 28(5):749–771, May 2002.

[199] M. Antonioletti, M. Atkinson, R. Baxter, A. Borley, N. P. C. Hong, B. Collins, N. Hardman, A. Hume, A. Knox, M. Jackson, A. Krause, S. Laws, J. Magowan, N. W. Paton, D. Pearson, T. Sugden, P. Watson, and M. Westhead. The Design and Implementation of Grid Database Services in OGSA-DAI. *Concurrency and Computation: Practice and Experience*, 17(2–4):357–376, Feb. 2005.

[200] J. Bester, I. Foster, C. Kesselman, J. Tedesco, and S. Tuecke. GASS: A Data Movement and Access Service for Wide Area Computing Systems. In *Proceedings of the 6th Workshop on I/O in Parallel and Distributed Systems*, pages 78–88, Atlanta, Georgia, May 1999. ACM Press.

[201] R. Bhattacharyya. Java FTP server. Available online at: `http://www.myjavaserver.com/~ranab/ftp` (Accessed August 31st, 2007).

[202] F. Cappello, S. Djilali, G. Fedak, T. Hérault, F. Magniette, V. Néri, and O. Lodygensky. Computing on large-scale distributed systems: Xtremweb architecture, programming models, security, tests and convergence with grid. *Future Generation Computer System*, 21(3):417–437, 2005.

[203] A. Chervenak, I. Foster, C. Kesselman, C. Salisbury, and S. Tuecke. The Data Grid: Towards an Architecture for the Distributed Management and Analysis of Large Scientific Datasets. *Journal of Network and Computer Applications*, 23:187–200, 1999.

[204] G. Fedak, C. Germain, V. Neri, and F. Cappello. XtremWeb: A Generic Global Computing System. In *CCGRID '01: Proceedings of the 1st International Symposium on Cluster Computing and the Grid*, pages 582–588. IEEE Computer Society, May 2001.

[205] R. Fielding, J. Gettys, J. Mogul, H. Frystyk, and T. Berners-Lee. Enabling Grids for E-sciencE (EGEE), 2006. Available online at: `http://www.eu-egee.org` (Accessed August 31st, 2007).

[206] R. Fielding, U. Irvine, J. Gettys, J. Mogul, H. Frystyk, and T. Berners-Lee. RFC-2068: Hypertext Transfer Protocol - HTTP/1.1, 1997. Available online at: `http://www.w3.org/Protocols/rfc2068/rfc2068` (Accessed August 31st, 2007).

[207] R. J. O. Figueiredo, N. H. Kapadia, and J. A. B. Fortes. The PUNCH Virtual File System: Seamless Access to Decentralized Storage Services in a Computational Grid. In *Proceedings of the 10th IEEE International Symposium on High Performance Distributed Computing (HPDC'01)*, pages 334–344, San Francisco, CA, Aug. 2001. IEEE, IEEE Press.

[208] FileZilla. Available online at: `http://filezilla.sourceforge.net` (Accessed August 31st, 2007).

[209] I. Foster. Globus Toolkit Version 4: Software for Service-Oriented Systems. In *IFIP International Conference on Network and Parallel Computing*, volume 3779 of *Lecture NOTEs in Computer Science*, pages 2–13. Springer-Verlag, 2005.

[210] I. Foster, C. Kesselman, G. Tsudik, and S. Tuecke. A Security Architecture for Computational Grids. In *Proceedings of the 5th ACM Conference on Computer and Communications Security*, pages 83–92, San Francisco, California, Nov. 2-5 1998. ACM Press.

[211] I. Foster, C. Kesselman, and S. Tuecke. The Anatomy of the Grid: Enabling Scalable Virtual Organizations. *The International Journal of High Performance Computing Applications*, 15(3):200–222, 2001.

[212] F. Garcia-Carballeira, J. Carretero, A. Calderón, J. D. Garcia, and L. M. Sanchez. A Global and Parallel File System for Grids. *Future Generation Computer Systems*, 23(1):116–122, Jan. 2007.

[213] GGF Grid File System working group (gfs-wg). Available online at: `https://forge.gridforum.org/projects/gfs-wg` (Accessed August 31st, 2007).

[214] GGF Grid File System working group (gfs-wg). The GGF Grid File System architecture workbook, Jan. 2006. Available online at: `http://www.ggf.org/documents/GFD.61.pdf` (Accessed August 31st, 2007).

[215] Globus. The Globus Resource Specification Language RSL v1.0, 2000. Available online at: `http://www.globus.org/toolkit/docs/2.4/gram/rsl_spec1.html` (Accessed August 31st, 2007).

[216] Globus project. Available online at: `http://www.globus.org` (Accessed August 31st, 2007).

[217] J. Howard, M. Kazar, S. Menees, D. Nichols, M. Satyanarayanan, R. Sidebotham, and M. West. Scale and Performance in a Distributed File System. *ACM Transactions on Computer Systems*, 6(1):51–81, Feb. 1998.

[218] JSCH - Java Secure Channel. Available online at: `http://www.jcraft.com/jsch` (Accessed August 31st, 2007).

[219] T. Kosar and M. Livny. A framework for reliable and efficient data placement in distributed computing systems. *Journal of Parallel and Distributed Computing*, 65(10):1146–1157, 2005.

[220] P. Kunszt and P. Badino. EGEE gLite User's Guide - Overview of gLite Data Management. Technical report egee-tech-570643-v1.0, CERN, Geneva, Switzerland, 2005.

[221] R. K. Madduri, C. S. Hood, and W. E. Allcock. Reliable File Transfer in Grid Environments. In *LCN '02: Proceedings of the 27th Annual IEEE Conference on Local Computer Networks*, pages 737–738, Washington, DC, 2002. IEEE Computer Society.

[222] T.-M.-H. Nguyen and F. Magoulès. A framework for data management in the grid. In *Proceedings of International Conference on Distributed Computing and Applications for Business, Engineering and Sciences*, pages 629–633, YiChang, Hubei, China, Aug. 2007. Hubei Science and Technology Press.

[223] T.-M.-H. Nguyen, F. Magoulès, and C. Révillon. GRAVY: Towards virtual file system for the grid. In *Proceedings of Advances in Grid and Pervasive Computing*, pages 567–578, Paris, France, May 2007. Springer Verlag.

[224] M. Pereira, O. Tatebe, L. Luan, and T. Anderson. Resource Namespace Service specification, May 2006. Available online at: `http://www.ggf.org/GGF17/materials/272/Resource_Namespace_Service_Refactored.pdf` (Accessed August 31st, 2007).

[225] J. Postel and J. Reynolds. RFC-959: File Transfer Protocol. Available online at: `http://www.w3.org/Protocols/rfc959/` (Accessed August 31st, 2007).

[226] SecureFTP. Available online at: `http://www.glub.com/products/secureftp` (Accessed August 31st, 2007).

[227] SmartFTP. Available online at: `http://www.smartftp.com` (Accessed August 31st, 2007).

[228] B. Wei, G. Fedak, and F. Cappello. Collaborative Data Distribution with BitTorrent for Computational Desktop Grids. In *ISPDC '05: Proceedings of the The 4th International Symposium on Parallel and Distributed Computing (ISPDC'05)*, pages 250–257, Washington, DC, 2005. IEEE Computer Society.

[229] B. Wei, G. Fedak, and F. Cappello. Scheduling Independent Tasks Sharing Large Data Distributed with BitTorrent. In *Proceedings of Grid Computing, 2005. The 6th IEEE/ACM International Workshop on*, 2005.

[230] B. S. White, M. Walker, M. Humphrey, and A. Grimshaw. LegionFS: A Secure and Scalable File System Supporting Cross-Domain High-Performance Applications. In *Proceedings of the IEEE/ACM Super-computing Conference (SC2001)*, pages 59–59, Denver, Colorado, Nov. 2001.

[231] A. Woehrer, P. Brezany, and I. Janciak. Virtualization of Heterogeneous Data Sources for Grid Information Systems. In *Proceedings of MIPRO 2004*, Opatija, Croatia, 24–28 may 2004.

[232] G. F. R. D. A. working group. GridFTP: Protocol extensions to FTP for the Grid, 2000. Available online at: `http://www.ggf.org/documents/GFD.47.pdf` (Accessed August 31st, 2007).

[233] T. Ylonen and C. Lonvick. RFC-4251: The Secure Shell (SSH) Protocol. Available online at: `http://www.ietf.org/rfc/rfc4251.txt` (Accessed August 31st, 2007).

Chapter 6

Scheduling grid services

6.1 Introduction

In the past twenty years, the parallel and distributed computing have been widely researched and utilized in the industry and scientific research. The dramatic growth in the number of powerful, easy-to-use, portable, and affordable computers, combined with globally accessible communication networks, has resulted in a large and growing user community which demands the sophisticated computing and services. In order to provide reliable and fast distributed services and to reduce the turn-around time of user jobs, the *scheduling algorithms* and strategies have been heavily studied. In the same time, a lot of systems (e.g. *Condor*, *PBS*) which provide the job queuing mechanism, scheduling policy and local resource management are developed to facilitate the creation and utilization of powerful clusters.

Along with the deployment of more and more heterogeneous clusters, the problem of requiring middleware to leverage existing IT infrastructure to optimize compute resources and manage data and computing workloads has emerged. Grid computing has become an increasingly popular solution to optimize resource allocation and integrate variable computing resources in highly charged IT environments.

Grid technologies and infrastructures support the integration and coordinated use of diverse resources in dynamic, distributed virtual organizations. According to the definition of Ian Foster [257], the characteristics of the Grid can be concluded in three points:

- coordinates resources that live within different control domains (e.g. different administrative units of the same or different companies). Thus the administration of each resource is independent and distributed.

- uses multi-purpose protocols and interfaces that address such fundamental issues as authentication, authorization, resource discovery, and resource access. But it is important that these protocols and interfaces be standard and open.

- allows its integral resources to be used in a coordinated fashion to deliver various qualities of service.

Thus new scheduling algorithms and strategies must be researched to take into account the characteristic issues of grids. In a grid environment, the scheduling problem is to schedule a stream of applications from different users to a set of computing resources to maximize system utilization. In the same time, *SOA*(Service-Oriented Architecture) is more and more adopted in industry and business domains as a common and effective solution to resolve the grid computing problem. Service-orientation describes an architecture that uses loosely coupled services to support the requirements of business processes and users. Resources in a *SOA* environment are made available as independent services that can be accessed without knowledge of their underlying platform implementation. Therefore efficient discovery of grid services is essential for the success of grid computing. In this chapter, the principal components of the grid scheduling will be presented, such as the service discovery, resource information and grid scheduling architecture. Fault-tolerance, the most important component to assure the qualities of service in a grid, is described at the end of this chapter.

6.2   Scheduling algorithms and strategies

A schedule of tasks (or schedule) is the assignment of tasks to specific time intervals of resources, such that no two tasks are on any resource at the same time, or such that the capacity of the resource is not exceeded by the tasks [255]. In general, *heterogeneous computing (HC)* is the coordinated use of different types of machines, networks, and interfaces to maximize their combined performance and/or cost-effectiveness [253, 260]. HC is an important technique for efficiently solving collections of computationally intensive problems [258]. Thus scheduling algorithms on heterogeneous computing systems have been widely studied.

Two types of scheduling algorithms are intensively researched: static and dynamic. In a general HC system, schemes are necessary to assign tasks to machines (matching) and to compute the execution order of the tasks assigned to each machine (scheduling) [242]. The process of matching and scheduling tasks is referred to as mapping. Dynamic methods perform the mapping as tasks arrive. This is in contrast to static techniques, where the complete set of tasks to be mapped is known a priori, the mapping is done prior to the execution of any of the tasks, and more time is available to compute the mapping.

6.2.1   Static heuristics

In order to clearly describe the heuristics, some preliminary terms must be defined. *Machine Availability Time*, $mat(j)$, is the earliest time a machine j

can complete the execution of all the tasks that have previously been assigned to it. Let $ETC(i,j)$ be defined as the estimated time to compute for task i on machine j. *Completion Time*, $ct(i,j)$, is the machine availability time plus the execution time of task i on machine j, $ct(i,j) = mat(j) + ETC(i,j)$. Let t as the number of tasks to be executed and m as the number of machines in the HC system. The maximum value of $ct(i,j)$, for $0 \leq i < t$ and $0 \leq j < m$ is known as the *makespan*. Each heuristic is attempting to minimize the makespan [241]. According to the description of papers [237, 241, 288, 279],we conclude heuristics as follows:

Opportunistic Load Balancing (OLB): assigns each task, in arbitrary order, to the next available machine, regardless of the task's estimated execution time on that machine.

User-Directed Assignment (UDA): in contrast to OLB, UDA assigns each task, in arbitrary order, to the machine with the best estimated execution time for that task, regardless of that machine's availability.

Fast Greedy: assigns each task, in arbitrary order, to the machine with the minimum completion time for that task.

Min-min: begins with the set U of all unmapped tasks. Then the set of minimum completion times,

$$M = \{m_i : m_i = min_{0 \leq j < m}(ct(i,j)), for \ \ each \ \ i \in U\},$$

is found. Next, the task with the overall minimum completion time from M is selected and assigned to the corresponding machine. Lastly, the newly mapped task is removed from U and the process repeats until all tasks are mapped.

Max-min: is very similar to Min-min. The Max-min also begins with the set U of all unmapped tasks. Then the set of minimum completion times,

$$M = \{m_i : m_i = min_{0 \leq j < m}(ct(i,j)), for \ \ each \ \ i \in U\},$$

is found. Next, the task with the overall maximum completion time from M is selected and assigned to the corresponding machine. Lastly, the newly mapped task is removed from U and the process repeats until all tasks are mapped.

Greedy: is literally a combination of the Min-min and Max-min heuristics. The Greedy heuristic performs both of the Min-min and Max-min heuristics, and uses the better solution.

Genetic Algorithm (GA): is a promising heuristic approach to find near-optimal solutions in large search spaces. The first step necessary to employ a GA is to initiate a population of chromosomes (possible mappings). A random set of chromosomes is often used as the initial population. This initial population is the first generation from which the evolution starts. Then all of chromosomes in the population are evaluated based on their *fitness value* (i.e., *makespan*). Thus, in this research a smaller fitness value represents a better solution. The selection process is the next step. In this step, each chromosome is eliminated or duplicated (one or more times) based on its relative quality. The population size is typically kept constant. Selection is followed by the crossover step. With some probability, some pairs of chromosomes are selected from the current population and some of their corresponding components are exchanged to form two valid chromosomes, which may or may not already be in the current population. After crossover, each chromosome in the population may be mutated with some probability. The mutation process transforms a chromosome into another valid one that may or may not already be in the current population. The new population is then evaluated. If the stopping criteria have not been met, the new population goes through another cycle (iteration) of selection, crossover, mutation, and evaluation. These cycles continue until one of the stopping criteria is met.

Simulated Annealing (SA): uses a procedure that probabilistically allows poorer solutions to be accepted to obtain a better result. This probability is based on a system temperature that decreases for each iteration. The initial system temperature is the *makespan* of the initial mapping and the initial mapping is generated and mutated in the same manner as the GA. Then the makespan is evaluated. If the new makespan is better, the new mapping replaces the old one. If the new makespan is worse, a function, $P(makespan_{new}, makespan_{old}, temperature)$, will be used to decide to accept or reject the new makespan.

Genetic Simulated Annealing (GSA): is a combination of the GA and SA techniques. In general, GSA follows procedures similar to the GA outlined above. However, for the selection process, GSA uses the SA cooling schedule and system temperature, and a simplified SA decision process for accepting or rejecting new chromosomes.

Tabu: The Tabu search keeps track of the regions of the solution space which have already been searched so as not to repeat a search near these 'Tabu' areas. A solution (mapping) uses the same representation as a chromosome in the GA approach. Heuristic searches are conducted within a region, and the best solution for that region is stored. Then, a new region, not on the tabu list, is searched. When a stopping criterion is reached, the best solution among regions is selected.

A*: is a tree-based search that has been applied to many task allocation problems. As the tree grows, intermediate nodes represent partial solutions

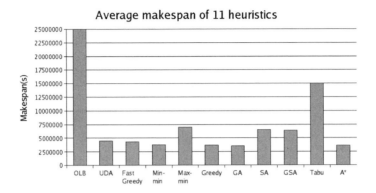

FIGURE 6.1: Inconsistent, high task, high machine heterogeneity.

(a subset of tasks are assigned to machines), and leaf nodes represent final solutions (all tasks are assigned to machines). The partial solution of a child node has one more task a mapped than the parent node. Each parent node can be replaced by its m children, one for each possible mapping of a. The number of nodes allowed in the tree is bounded to limit mapper execution time. For each node, n, we can associate a cost function, $f(n)$, with it. $f(n)$ represents the makespan of the partial solution of node n plus a lower-bound estimate of the time to execute the unmapped tasks in the meta-task. Thus the node with the minimum $f(n)$ is replaced by its m children and the tree is expanded. If the number of created nodes reaches the maximum number of allowed nodes, the node with the largest $f(n)$ is deleted. The process continues until a leaf node (complete mapping) is reached.

Figures 6.1 and 6.2 show comparisons of the 11 static heuristics using *makespan* as the criterion in two dierent heterogeneity environments. For each heuristic, there are 512 tasks which are submitted to 16 machines and 100 trials are executed. The bars in the figures show the averages of *makespan*. It can be seen that, for the parameters used in this study, GA gives the smallest makespan for inconsistent heterogeneities [279].

6.2.2 Dynamic heuristics

In an HC system where the tasks to be executed are not known a priori, dynamic schemes are necessary to match tasks to machines, and to compute the execution order of the tasks assigned to each machine [279]. The mapping heuristics can be grouped into two categories: on-line mode and batch-mode heuristics. In the on-line mode, a task is mapped onto a machine as soon as it arrives at the mapper. In the batch mode, tasks are not mapped onto the machines as they arrive; instead they are collected into a set that is examined

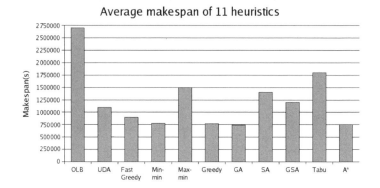

FIGURE 6.2: Inconsistent, high task, low machine heterogeneity.

for mapping at prescheduled times called mapping events.

6.2.2.1   On-line mode heuristics

The *Minimum Completion Time (MCT) heuristic* assigns each task to the machine that results in that task's earliest completion time. This causes some tasks to be assigned to machines that do not have the minimum execution time for them. As a task arrives, all the machines in the HC suite are examined to determine the machine that gives the earliest completion time for the task [273].

The *Minimum Execution Time (MET) heuristic* assigns each task to the machine that performs that task's computation in the least amount of execution time. This heuristic, in contrast to the MCT, does not consider machine ready times, and can cause a severe imbalance in load across the machines. The main advantage of this method is its simplicity [279].

The *Switching Algorithm (SA) heuristic* is motivated by the following observation. The MET heuristic can potentially create load imbalance across machines by assigning many more tasks to some machines than to others, whereas the MCT heuristic tries to balance the load by assigning tasks for earliest completion time. The SA heuristic uses the MCT and MET heuristics in a cyclic fashion depending on the load distribution across the machines. The purpose is to have a heuristic with the desirable properties of both the MCT and the MET [279].

The *KPB (K-Percent Best) heuristic* considers only a subset of machines while mapping a task. The subset is formed by picking the $(k \times m/100)$ best machines based on the execution times for the task, where $100/m \leq k \leq 100$. The task is assigned to a machine that provides the earliest completion time in the subset. If $k = 100$, then the KPB heuristic is reduced to the MCT heuristic. If $k = 100/m$, then the KPB heuristic is reduced to the MET heuristic. A 'good' value of k maps a task to a machine only within a subset formed from machines computationally superior for that particular task [273].

The *Opportunistic Load Balancing (OLB) heuristic* assigns a task to the machine that becomes ready next, without considering the execution time of the task on that machine. If multiple machines become ready at the same time, then one machine is arbitrarily chosen. The complexity of the OLB heuristic is dependent on the implementation. In the implementation considered here, the mapper may need to examine all m machines to find the machine that becomes ready next. Therefore, it takes O(m) to find the assignment. Other implementations may require idle machines to assign tasks to themselves by accessing a shared global queue of tasks [282].

6.2.2.2 Batch mode heuristics

In the batch mode heuristics, meta-tasks are mapped after predefined intervals. These intervals are defined using one of the two strategies proposed below [273].

1. The regular time interval strategy maps the meta-tasks at regular intervals of time (e.g., every 10 sec).

2. The fixed count strategy. In this strategy, the length of the mapping intervals will depend on the arrival rate and the completion rate.

The *Min-min heuristic* is shown in Algorithm 6.2.1. It calculates the earliest completion time for each task, t_j in the meta-task M and the machine, m_i, that obtains it. Then it finds the task, t_k, with the minimum earliest completion time in M. Next, the task, t_k is assigned to the machine that gives the earliest completion time of the task, and that machine's ready time is updated. This assigned task is removed from the meta-task and the procedure is repeated [273].

The *Max-min heuristic* is similar to the Min-min heuristic. Once it finds all the earliest completion time of each task in the meta-task, the task t_k that has the maximum earliest completion time is determined and then assigned to the corresponding machine. Then this assigned task is removed from the meta-task and this procedure is repeated until all of the tasks have been mapped. [273].

The *Sufferage heuristic* (shown in Algorithm 6.2.2) finds the earliest completion time of each task in the meta-task and calculates the sufferage value of a task t_k. The sufferage value of tasks is the difference between its second earliest completion time (on some machine m_y) and its earliest completion time (on some machine m_x). For the task t_k, if the machine m_j that gives the earliest completion time is unassigned, the task t_k is assigned to that machine and is removed from the meta-task. Otherwise, the heuristic compares the sufferage value of the task t_i assigned in the machine with the sufferage value of task t_k. If the sufferage value of t_k is bigger than the value of task t_i, the task t_i is unassigned and added to the meta-task. The task t_k is assigned to

---

Algorithm 6.2.1 The Min-min heuristic

> **for** Each Task t_j in meta-task M **do**
>   **for** Each machine m_i **do**
>   Calculate ct(i,j) = mat(j) + ETC(i,j)
>   **end for**
> **end for**
> **while** Unmapped tasks remaining **do**
>   **for** Each task in M **do**
>   find its earliest completion time and the machine that obtains it
>   find the task t_k with the minimum earliest completion time
>   assign the task t_k to the machine that gives the earliest completion time
>   delete the task t_k from M
>   update
>   **end for**
> **end while**

---

that machine and is removed from the meta-task. Each task in the meta-task is considered only once [279].

For many heuristics, there are control parameter values and/or control function specifications that can be selected for a given implementation and such values and specifications are selected based on experimentation and/or information in the literature. Normally, the KPB provides the minimum makespan in the on-line mode heuristics and the Sufferage heuristic gives the smallest makespan in the batch mode heuristics [279].

6.2.3   Grid scheduling algorithms and strategies

The traditional parallel scheduling problem is to schedule the subtasks of an application to the parallel machines in order to reduce the *turn-around time* . In a grid environment, the scheduling problem is to schedule a stream of applications from different users to a set of computing resources to maximize system utilization. Both static and dynamic heuristics are widely adopted in grid computing. Dynamic scheduling is more appropriate than static scheduling in a grid environment because of the multiplicities of machines (e.g., processor speed).

In the dynamic scheduling, if a resource is assigned many tasks, it may invoke a balancing strategy to decide which task should be execute first. According to how the dynamic task assignment is achieved, there are two basic approaches [280]:

- *First-Come-First-Served (FCFS) policy.* The task is executed according to the assigned time. The task which has the earliest assigned time is executed first.

Algorithm 6.2.2 The Sufferage heuristic

---

for Each Task t_j in meta-task M **do**
  for Each machine m_i **do**
    Calculate ct(i,j) = mat(j) + ETC(i,j)
  end for
end for
while Unmapped tasks remaining **do**
  mark all machines as unassigned
  for each task t_k in M **do**
    find machine m_j that gives the earliest completion time
    sufferage value = second earliest completion time - earliest completion time
    if machine m_j is unassigned **then**
      assign t_k to machine m_j
      delete t_k from M
      mark m_j as assigned
    else
      if the sufferage value of t_i < the sufferage value of t_k **then** {the task is assigned to machine m_j is t_i}
        unassign t_i, add to M
        assign t_k to machine m_j, remove from M
      end if
    end if
  end for
  update
end while

- *Backfilling policy.* Backfilling works by identifying "holes" in the local job queue and moving forward smaller jobs that fit those holes. There are two common variations to backfilling - conservative and aggressive. In conservative backfill, every job is given a reservation when it enters the system. A smaller job is moved forward in the queue as long as it does not delay any previously queued job. In aggressive backfilling, only the job at the head of the queue has a reservation. A smaller job is allowed to leap forward as long as it does not delay the job at the head of the queue.

The economic approach for managing resource allocation in grid computing environments provides a fair basis in successfully managing decentralization and heterogeneity that is present in human economies [245]. There are two key players in the economic model: *Grid service providers (GSPs)* providing the traditional role of producers and *Grid resource brokers (GRBs)* representing consumers. Consumers interact with their own brokers for managing and scheduling their computations on the grid. The GSPs make their resources grid enabled by running software systems along with grid trading services to enable resource trading and execution of consumer requests directed through GRBs. GRBs may invite bids from a number of GSPs and select those that offer the lowest service costs and meet their deadline and budget requirements. Alternatively, GSPs may invite bids in an auction and offer services to the highest bidder as long as its objectives are met. Both GSPs and GRBs have their own utility functions that must be satisfied and maximized.

In [264], a *QoS* Guided Min-min heuristic is presented which can guarantee the QoS requirements of particular tasks and minimize the makespan at the same time. It divides all the tasks in the meta-task into two parts: tasks with high QoS request and tasks with low QoS request. The tasks with high QoS request will be mapped first to satisfy the QoS requirement. In each part, the Min-min heuristic is used to assign a task to the corresponding machine.

6.3   Architecture

The management of batch jobs within a single distributed system or domain has been addressed by many research and commercial systems, notably *Condor* [271], *LSF* [290], and *PBS* [266]. In a grid environment, the management of jobs on a set of heterogeneous, dynamically changing resources is a more complex problem. A meta-scheduler is a manager or supervisor of local resource managers, which control the use of individual resources such as clusters, computing farms, servers or supercomputers. The meta-scheduler is therefore a key component of a computational grid as it is responsible for optimizing the use of grid resources. A number of schedulers for grid computing systems have been developed.

6.3.1 Meta-schedulers

Condor-G [261] is an innovative distributed computing framework that addresses the management of computation and harnessing of resources, credential management, resource discovery and fault-tolerance. In brief, Condor-G combines the inter-domain resource management protocols of the *Globus Toolkit* [259] and the intra-domain resource management methods of Condor [271] to allow the user to harness multi-domain resources as if they all belong to one personal domain. The advantages of Condor-G are shown below.

- It allows the user to treat the grid as an entirely local resource, with an *API* and command line tools that allow the user to submit jobs, to query a job's status, to be informed of job terminations or problems and to obtain access to detailed logs.

- Condor-G is built to tolerate four types of failure: crash of the Globus JobManager, crash of the machine that manages the remote resource, crash of the machine on which the GridManager is executing, and failures in the network connecting the two machines.

- A user-supplied list of GRAM servers or a personal resource broker are used to achieve the resource discovery and scheduling. The information from resources is gathered and then the Matchmaker is used to make brokering decisions.

- Credential management. Condor-G deals with credential expiration by periodically analyzing the credentials for all users with currently queued jobs. Credentials may have been forwarded to a remote location, in which case the remote credentials need to be refreshed as well.

Nimrod-G [244] is a grid resource broker that uses a computational economy driven architecture for managing resources and scheduling task farming applications on large-scale distributed resources. Its key components are: Client or User Station, Parametric Engine, Scheduler, Dispatcher and Job-Wrapper. The client or user station acts as a user-interface for controlling and supervising a job under consideration. The parametric engine is responsible for managing the execution of parametrized application jobs. It takes care of the actual creation of jobs, the maintenance of job status, and providing a means for interaction between the clients, the schedule advisor, and the dispatcher. The scheduler is responsible for resource discovery, resource trading, resource selection, and job assignment. The scheduler can use the information gathered by a resource discoverer and also negotiate with resource owners to establish service price. The resource that offers the best price and meets resource requirements can eventually be selected. The dispatcher primarily initiates the execution of a task on the selected resource as per the scheduler's instruction. The job-wrapper is responsible for staging application tasks and data; starting execution of the task on the assigned resource and sending results back to the parametric engine via dispatcher.

GrADS [250] aims to produce a software execution environment for code to be run on a computational grid and is designed specifically for Parameter Sweep Application. Two improvements of the GrADS Project can be mentioned:

- Rescheduling by stop/migration/restart and by single-processor swapping are both feasible, flexible and require little additional programming.

- A new GrADS workflow scheduler that resolves the application dependence and schedules the components, including parallel components, onto available resources is developed. For each application component, the GrADS workflow scheduler ranks each eligible resource, reflecting the fit between the component and the resource. Then the scheduler collates this information into a performance matrix. Finally, it runs heuristics on the performance matrix to schedule components onto resources.

gLite [243] is born from the collaborative efforts of academic and industrial research centers as part of the *EGEE* [252] Project. gLite consists of six principal components: Computing Element, Workload Management, Storage Element, Catalog, Information and Monitoring and Security. It achieves efficiently and reliably the scheduling of computational tasks on the available infrastructure, the data storage and movement on the infrastructure, the provision of grid information and application monitoring data. The workload management system in *gLite* supports more advanced job types: Normal, DAG, MPI, Checkpointable and Interactive. The LB (Logging and Bookkeeping) service is used to keep track of a job's status. A system called VOMS (Virtual Organization Membership Service) is used to manage information about the roles and privileges of users within a VO. This information is presented to services via an extension to the proxy.

GridWay [263] is an open source meta-scheduling technology that performs job execution management and resource brokering on heterogeneous and dynamic grids based on Globus Toolkit services. GridWay allows unattended, reliable, and efficient execution of single, array, or complex jobs on heterogeneous and dynamic grids. GridWay performs all the job scheduling and submission steps transparently to the end user and adapts job execution to changing grid conditions by providing fault recovery mechanisms, dynamic scheduling, migration on-request and opportunistic migration. GridWay on Globus provides decoupling between applications and the underlying local management systems. It provides full support for C and JAVA DRMAA (Distributed Resource Management API) GGF standard for the development of distributed applications and a command line interface similar to that found in local resource managers, and its modular design allows an easy incorporation of new grid services and so inter-operability between different grid infrastructures (Globus WS, Globus pre-WS and EGEE).

6.3.2 Grid scheduling scenarios

In the past years, many grids have been developed and implemented in scientific research and industry domains. However, these grid systems provide only domain-specific solutions to the problem of scheduling resources in a grid and no common and generic grid scheduling system has emerged yet. Thus some generic features of the grid must be identified in order to define the prototype of a genetic grid scheduling system. Tonellotto, N., Yahyapour, R. and Wieder, P. [286] present three common grid scheduling scenarios in the grid.

Enterprise Grids Enterprise Grids represent a scenario of commercial interest in which the available IT resources within a company are better exploited and the administrative overhead is lowered by the employment of grid technologies. The resources are typically not owned by different providers and are therefore not part of different administrative domains. In this scenario a centralized scheduling architecture (i.e. a central broker) is the single access point to the whole infrastructure and interacts directly with the local resource managers. Every user must submit jobs to this centralized entity.

High Performance Computing Grids High Performance Computing Grids represent a scenario in which different computing sites (e.g. scientific research labs) collaborate for joint research. Computing resources that execute compute- and/or data-intensive applications are usually large parallel computers or cluster systems. In this case the resources are part of several administrative domains, with their own policies and rules. A user can submit jobs to the broker at institute or VO level. The brokers can split a scheduling problem into several sub-problems, or forward the whole problem to different brokers in the same VO.

Global Grids Global Grids might comprise all kinds of resources, from single desktop machines to large-scale HPC machines, which are connected through a global grid network. This scenario is the most general one, covering both cases illustrated above and introducing a fully decentralized architecture.

6.3.3 Metascheduling schemes

The hierarchy of the metascheduler and computing resources and the role which the metascheduler plays in the job scheduling can be defined as the metascheduling schemes.   Three metascheduling schemes are discussed in [281].

Centralized Scheme In the centralized model, the metascheduler maintains information about all sites. All jobs are submitted to the metascheduler. Based on the queue of jobs submitted, and the information about all the constituent sites, the metascheduler makes scheduling decisions. With this model, the local sites are responsible only for dispatching the jobs that are supplied by the metascheduler, and providing information to the metascheduler.

Hierarchical Scheme With the hierarchical scheme, all jobs are still submitted to the metascheduler. But unlike the centralized scheme, jobs are not maintained in the metascheduler queue until dispatch time. Each site maintains a local queue from which it schedules jobs for execution. It is possible for different sites to use different scheduling policies. Once submitted to a local scheduler, the metascheduler has no further direct influence on the scheduling of the job, and the job cannot be moved to another site even if the load at the other site becomes lower at some time in the future.

Distributed Scheme This scheme is similar to the hierarchical scheme except that there is a metascheduler at every site and jobs are submitted to the local metascheduler where the job originates. The metaschedulers query each other periodically to collect instantaneous load information. If any of the other schedulers has a lower load, the job is transferred to the site with the lowest load. Since all jobs are submitted locally, the distributed scheme is more scalable than the hierarchical scheme.

6.4   Service discovery

Efficient discovery of grid services is essential for the success of grid computing. The standardization of grids based on Web services has resulted in the need for scalable Web service discovery mechanisms to be deployed in grids [239]. There are two techniques to perform the process of services discovery: syntactic and semantic [274]. Such processes must rely on the storage of arbitrary metadata about services that originate from both service providers and service user. Several standards of service directory are studied to support the storage and the organization of the metadata.

6.4.1   Service directories

Early distributed systems comprise collections of components that are implicitly linked through function names, or linked through TCP/IP-based hosts and port addresses. Domain Name Servers and Jini [238] simplify and abstract the use of these implicit links by providing a registering mechanism for local components [274]. Grid computing is based on standards which use Web services technology. In the grid environment, the service discovery function is assigned to a specialized grid service called registry. The service directories that support the registry and the discovery of the grid services are known as:

6.4.1.1   UDDI

UDDI (Universal Description, Discovery, and Integration) is an XML-based registry for businesses worldwide to list themselves on the Internet. Its ultimate goal is to streamline online transactions by enabling companies to find one another on the Web and make their systems inter-operable for e-commerce. The project allows businesses to list themselves by name, product, location, or the Web services they offer. A UDDI registry enables a business to enter three types of information in a UDDI registry: white pages, yellow pages and green pages. UDDI's intent is to function as a registry for services just like in Yellow pages; companies register themselves and their services under different categories. In UDDI, White Pages are a listing of the business entities. Green pages represent the technical information that is necessary to invoke a given service. And Yellow pages give more details about the business entities in the White pages. Thus, by browsing a UDDI registry, a developer should be able to locate a service and a company and find out how to invoke the service.

However, today UDDI has not been widely deployed in the Internet because of its shortcomings. For example it does not provide the ability to query high-level service information such as identifying services within a particular price range. In fact, the only known uses of UDDI are what are known as private UDDI registries within an enterprise's boundaries. Improvement of the UDDI standard is continuing in full force and UDDI version 3 (V3) was recently approved as an OASIS Standard. However, UDDI today has issues that have not been addressed, such as scalability and autonomy of individual registries.

6.4.1.2   MDS

The Monitoring and Discovery System (*MDS*) [262] is a suite of Web services to monitor and discover resources and services on Grids. This system allows users to discover what resources are considered part of a Virtual Organization (VO) and to monitor those resources. MDS services provide query and subscription interfaces to arbitrarily detailed resource data and a trigger interface that can be configured to take action when pre-configured trouble conditions are met. The services included in the WS MDS implementation (MDS4), provided with the Globus Toolkit 4, acquire their information through an extensible interface which can be used to:

- query *WSRF* services for resource property information,

- execute a program to acquire data,

- interface with third-party monitoring systems.

MDS4 includes two WSRF-based services: an Index Service, which collects data from various sources and provides a query/subscription interface to that data, and a Trigger Service, which collects data from various sources and can be configured to take action based on that data. An Archive Service, which will provide access to historic data, is planned for a future release.

The Index Service is a registry similar to UDDI, but much more flexible. Indexes collect information and publish that information as resource properties. Clients use the standard WSRF resource property query and subscription/notification interfaces to retrieve information from an Index. Indexes can register to each other in a hierarchical fashion in order to aggregate data at several levels. Indexes are "self-cleaning"; each Index entry has a lifetime and will be removed from the Index if it is not refreshed before it expires.

Each Globus container that has MDS4 installed will automatically have a default Index Service instance. By default, any GRAM, RFT, or CAS service running in that container will register itself to the container's default Index Service.

6.4.1.3   ICENI

ICENI (Imperial College e-Science Networked Infrastructure), developed by the London e-Science Center, supports the concept of a computational community based on Jini technology [238] . The participants in the computational community publish their services through Jini lookup service, which serves as a registry in a Jini environment. Services can be identified through lookup services by matching data members of their entry object with requests. Compared to Web services based registry, Jini facilitates dynamic service discovery but restricts end-points in a pure Java environment [289].

6.4.2   Techniques syntactic and semantic

6.4.2.1   Syntactic service discovery

Syntactic service discovery mainly focuses on the abstract part of a WSDL description: operation and input/output messages. Service discovery is performed by querying the name or the type of a service. The service is advertised by its service information (i.e. name, type) in the registry. By retrieving this service information, the user can discover services.

The paper [239] presents a distributed Web-service discovery architecture, called *DUDE* (Distributed UDDI Deployment Engine). DUDE leverages *DHT* (Distributed Hash Tables) as a rendezvous mechanism between multiple UDDI registries. DUDE enables consumers to query multiple registries, still at the same time allowing organizations to have autonomous control over their registries. The DUDE architecture can support effective distribution of UDDI registries thereby making UDDI more robust and also addressing its scaling issues. Furthermore, the DUDE architecture for scalable distribution can be applied beyond UDDI to any grid service discovery mechanism.

A Grid Market Directory (GMD) system is proposed in [289]. The GMD is developed as a VO marketplace registry, which serves as a registry for publication and discovery of grid service providers and their services. The GMD consists of two key components: the portal manager and the query Web-service. The GMD portal manager is responsible for provider registration, service publication and management, and service browsing. All these tasks are accomplished by using a standard web browser. The GMD query Web-service provides services so that clients such as resource brokers can query the GMD and obtain the information of resources to discover those that satisfy the user QoS requirements.

The Web Service Discovery Architecture (WSDA), proposed by Hoschek [267], specifies communication primitives useful for discovery, service identification, service description retrieval, data publication as well as query support. The individual primitives can be combined and plugged together by specific clients and services to yield a wide range of behaviors. A hyper registry is also introduced which is a centralized database node for discovery of dynamic distributed content. This registry supports XQueries over a tuple set from a dynamic XML data model. The architecture includes a Unified Peer-to-Peer Database Framework (UPDF) and a corresponding Peer Database Protocol (PDP) [236].

6.4.2.2 Semantic service discovery

Semantic service discovery takes into account the semantic meaning of a parameter in addition to syntactic matching. The interface description is transformed into an ontology which is a knowledge schema especially for services. Semantic information of services consists of their extensive descriptions including, but not limited to, capabilities, functionality, portability and system requirements. Semantic service matching introduces the possibilities of fuzziness and inexactness of the response to a service discovery request.

The paper [270] proposes a flexible ontology management approach for discovery and description of grid service capabilities supporting ontology evolution whose goal is to enhance the inter-operability among grid services. In a domain, a concept may have changed or have emerged. The ability to be able to reflect this change in a ontology is called ontology evolution. In this approach, concepts and descriptions in an ontology are defined independently, and they are connected by relationships. In addition, the relationships are updated based on real-time evaluations of ontology users in order to flexibly support ontology evolution. A bottom-up ontology evolution means such an environment that allows ontology users to evaluate impact factors of concepts in an ontology and that results of the evaluation are reflected in the modification of the ontology. The contribution of this paper is to suggest the ontology management framework that not only enables semantic discovery and description of a grid service capability but also supports a bottom-up ontology evolution based on the users' evaluations.

A framework for ontology-based flexible discovery of Semantic Web services is described in [276]. The proposed approach relies on user-supplied, context-specific mappings from a user ontology to relevant domain ontologies used to specify Web services. The mechanism that transforms a user's query for a Web service into queries and the process of matchmaking engine are presented. The framework also describes how user-specified preferences for Web services in terms of non-functional requirements (e.g., QoS) can be incorporated into the Web service discovery mechanism to generate a partially ordered list of services that meet user-specified functional requirements.

S. A. Ludwig and P. van Santen [272] propose a service discovery matchmaking framework based on a well-defined ontology. The matching mechanism comprises three filter stages. These are context, syntactic and semantic matching, whereas the service ontology database provides the knowledge-base. To allow matching engines to perform flexible matches, service requesters are allowed to decide the degree of flexibility that they grant to the system. Semantic matching is based on DAML ontologies. The advertisements and requests refer to DAML concepts and the associated semantic. Using this matchmaking framework allows for a better service discovery and close matches in a flexible way based on the defined ontology.

The project DReggie [248] presents a dynamic service discovery infrastructure that uses DAML to describe services and a Prolog reasoning engine to perform matching using the semantic content of service descriptions. At the heart of DReggie is an enhanced Jini Lookup Service (JLS) that enables smart discovery of Jini-enabled services. This infrastructure should be a necessary component of mobile devices and wireless networks.

6.5   Resource information

High-performance execution in distributed computing environments often requires careful selection and configuration not only of computers, networks, and other resources but also of the protocols and algorithms used by applications. Selection and configuration in turn require access to accurate, up-to-date information on the structure and state of available resources [256]. In the grid, information services which can perform the discovery, characterization, and monitoring of resources, services, and computations are a vital part of any grid software infrastructure.

6.5.1 Globus Toolkit information service

Globus Monitoring and Data Service (MDS2) has been used in many grid systems. It is now deprecated by Globus Alliance. The initial implementation used the GRIS (Grid Resource Information Service) and the GIIS (Grid Index Information Service). This was a distributed service with one GRIS per resource and one GIIS per site. In all its various implementations it suffered from scalability and stability problems [285].

Globus MDS4 [262] includes an Aggregator Framework, which provides a unified mechanism used by Services (such as the Index and Trigger services) built on it. Services built on the Aggregator Framework collect information via Aggregator Sources, a Java class that implements an interface (defined as part of the Aggregator Framework) to collect XML-formatted data. MDS4 includes the following three Aggregator Sources:

- the Query Aggregator Source, which polls a WSRF service for resource property information,

- the Subscription Aggregator Source, which collects data from a WSRF service via WSRF subscription/notification,

- the Execution Aggregator Source, which executes an administrator-supplied program to collect information.

Depending on the implementation, an Aggregator Source may use an external software component or a WSRF service may use an external component to create and update its resource properties. This set of components is called Information Providers. MDS4 supports the following information providers:

- *Hawkeye* Information Provider: An Information Provider that gathers Hawkeye data about Condor pool resources using the XML mapping of the GLUE schema and reports it to a WS GRAM service, which publishes it as resource properties.

- *Ganglia* Information Provider: An Information Provider that gathers cluster data from resources running Ganglia using the XML mapping of the GLUE schema and reports it to a WS GRAM service, which publishes it as resource properties.

- WS GRAM: The job submission service component of GT4. This WSRF service publishes information about the local scheduler.

- Reliable File Transfer Service (RFT): The file transfer service component of GT4.

- Community Authorization Service (CAS): This WSRF service publishes information identifying the virtual organization (VO) that it serves.

- Any other WSRF service that publishes resource properties.

6.5.2 Other information services and providers

The requirements of grid-based information systems are described in [251]. First, we require that a grid information service should focus only on efficient delivery of state information from a single source. If applications require accurate local state or consistent global state, this functionality can be achieved via other control functions that provide necessary atomic operations at a higher cost. Second, in distributed environments, both individual entities and the networks that provide access to those entities may fail. We hence require that information services behave robustly in the face of failure of any of the components on which the service is built. Finally, a new VO may involve many entities and have unique requirements for discovery and monitoring. We would like to be able to define once, ahead of time, the discovery and enquiry mechanisms that must be supported by any grid entity.

R-GMA [246] is part of gLite/EGEE and it is a monitoring and information management service for distributed resources. It exposes a relational model with *SQL* support to provide static as well as dynamic information about grid resources. The R-GMA architecture is based on that of the Grid Monitoring Architecture (GMA) [284] of the Global Grid Forum (GGF). The GMA consists of three components: consumers, producers and a directory service, which we refer to as a registry as it avoids any implied structure. In the GMA producers of information register themselves with the registry when they are instantiated. The registry, which may be distributed, describes the type and structure of information the producers want to make available to the grid. Potential consumers of information can query the registry to find out what type of information is available and locate producers that provide such information. Once a consumer has this information it can contact the producer directly to obtain the relevant data.

R-GMA can be used as a replacement for MDS. A small tool (GIN) has been written to invoke the MDS-like EDG info-providers and publish the information via R-GMA. R-GMA is also being used for network monitoring and to locate replica catalogs.

Grimoires (Grid RegIstry with Metadata Oriented Interface) [287] is a registry for the *myGrid* project [234] and the OMII Grid software release (www.omii.ac.uk). In Grimoires, a protocol for attaching metadata to registered service description, and for querying over service descriptions and their metadata is developed. The service registry is based on the UDDI (Universal Description Discovery and Integration) framework but extends the framework to allow metadata to be attached to various parts of the service description. An extensible programmatic interface is also provided for clients with an easy way to access the information, whether it is held in a remote Web service or locally. The result is an extremely flexible service registry that can be the basis of a semantically-enhanced service discovery engine [275].

NAREGI (National Research Grid Initiative) uses CIMOM, the CIM Object Manager,which distributes information about compute elements based on the Common Information Model, an emerging industry standard. This is then aggregated to a relational database and implemented as a grid service by use of Globus service OGSA DAI (Open Grid Services Architecture Data Access and Integration). Each site is referred to as a Cell Domain [285].

6.6 Data-intensive service scheduling

Data-intensive applications executing over a computational grid demand large data transfers which are normally costly operations. Therefore, taking them into account is mandatory to achieve efficient scheduling of data-intensive applications on grids.

6.6.1 Algorithms

As we have discussed in 6.2, there are many heuristics which are used to perform the scheduling of the grid. Although there are schedulers that attain good performance with these heuristics, they were not designed to take data transfer into account. The data-intensive application is the scientific or enterprise application that deals with a huge amount of data. Currently, there exist some algorithms that are able to take data transfer into account when scheduling data-intensive applications on grid environments [240][247][254][278].

- The paper [240] considers the problem of allocating a large number of independent, equal-sized tasks to a heterogeneous grid computing platform. Since the data transfer tasks can be split into subtasks and each subtask can often be processed independently, the solution presented in this paper can be used to deal with the data-intensive applications scheduling problem.

  A tree structure is used to model a grid, where resources can have different speeds of computation and communication, as well as different capabilities. The paper makes the assumption that the data for the computation initially resides at a single node of the tree, and the results of the computation will be returned to that same node. The node that serves as the source of the data is the root of the tree. In order to determine the allocation of tasks to nodes in the tree that maximizes the number of tasks executed per unit time in steady-state, the solution proposed is bandwidth-centric: tasks should be allocated to nodes in order of fastest communication time. If enough bandwidth is available, then all nodes are kept busy; if bandwidth is limited, then tasks should be allocated only to the children which have sufficiently small communication times, regardless of their computation power. The simulation results show that the bandwidth-centric method obtains better results than allocating tasks to all processors on a first-come, first served basis.

- The paper [254] presents a method, Adaptive Regression Method (AdRM), for predicting the performance of data transfer operations in network-bound distributed data-intensive applications. The Network Weather Service (NWS) is employed as network bandwidth probes to make short-term prediction of transfer time for a range of file sizes. AdRM combines NWS measurements with instrumentation data taken from actual application runs to predict the future performance of the application. The result is an accurate performance model that can be parameterized by "live" NWS measurements to make time-sensitive predictions.

  The scheduler incorporates both application-specific system requirements and dynamic resource performance information to schedule distributed applications in multi-user distributed environments. It uses a performance model based on the application's communication and computational needs. Performance models can be represented by mathematical equations, in which numeric values for resource performance forecasts are variables. The scheduler can evaluate the value of the performance equation for different resource mixes and choose the resource combination that maximizes application performance.

- An adaptive scheduling algorithm for data-intensive applications, XSufferage, is proposed in [247]. As we present in section 6.2, Sufferage can be used for scheduling independent tasks for a uniform single-user environment. This heuristic calculates the Minimum Completion Time (MCT) and a sufferage value which is defined as the difference between its best MCT and its second-best MCT for each task. Tasks with high sufferage value take precedence. However, the Sufferage heuristic does not lead to best makespans in some situations. Assume that a task, say T_0, requires a large input file that is already stored on a remote cluster. If that cluster contains two hosts with nearly identical performance, then both those hosts can achieve nearly the same MCT for that task. If the file is of significant size compared to network bandwidth available, then it is likely that those two hosts lead to the best and second-best MCTs for T_0. This means that the sufferage value will be close to zero, giving the task low priority. Other tasks may be scheduled in its place and force T_0 to be scheduled on some other clusters, thereby requiring an additional file transfer. This problem is solved in XSufferage by using a modified sufferage value definition. For each task and each cluster, we compute the task's MCT only for hosts in the given cluster and that value the cluster-level MCT. The cluster-level sufferage value is computed as the difference between the best and second-best cluster-level MCT. The task with the highest cluster-level sufferage value is given priority and is scheduled to the host that achieves the earliest MCT within the cluster.

- Ranganathan, K. and Foster, I. [277] describe a framework to address the large-scale data-intensive problems. Within this framework, data movement operations may be either tightly bound to job scheduling decisions or, alternatively, performed by a decoupled, asynchronous process on the basis of observed data access patterns and load. The scheduling logic is encapsulated in three modules:

  External Scheduler (ES): Users submit jobs to the External Scheduler they are associated with.

  Local Scheduler (LS): Once a job is assigned to run at a particular site (and sent to an incoming job queue) it is then managed by the Local Scheduler.

  Dataset Scheduler (DS): The DS at each site keeps track of the popularity of each dataset locally available. It then replicates popular datasets to remote sites depending on some algorithm.

  Most interestingly, they find that the framework can achieve particularly good performance with an approach in which jobs are always scheduled where data is located, and a separate replication process at each site periodically generates new replicas of popular datasets.

- The paper [278] presents Storage Affinity, a novel scheduling heuristic for bag-of-tasks data-intensive applications running on grid environments. Storage Affinity exploits a data reuse pattern, common on many data-intensive applications, that allows it to take data transfer delays into account and reduce the makespan of the application. Further, it uses a replication strategy that yields efficient schedules without relying upon dynamic information that is difficult to obtain.

  Storage Affinity was conceived to exploit data re-utilization to improve the performance of the application. Data re-utilization appears in two basic flavors: inter-job and inter-task. The former arises when a job uses the data already used by (or produced by) a job that executed previously, while the latter appears in applications whose tasks share the same input data. Thus, storage affinity of a task to a site is defined as the number of bytes within the task input dataset that are already stored in the site. The algorithm calculates the highest storage affinity value for each task. After this calculation, the task with the largest storage affinity value is chosen and scheduled. Since the information about data size and data location can be obtained a priori without difficulty and loss of accuracy, Storage Affinity does not use dynamic information about the grid and the application which is difficult to obtain. Storage Affinity applies also to task replication. Replicas have a chance to be submitted to faster processors than those processors assigned to the original task, thus increasing the chance of decreasing the task completion time.

6.6.2   Architecture of data grid

In domains as diverse as global climate change, high energy physics, and computational genomics, the volume of interesting data is already measured in terabytes and will soon total petabytes. The communities of researchers that need to access and analyze this data are often large and are almost always geographically distributed. But no integrating architecture exists that allows us to identify requirements and components common to different systems and hence apply different technologies in a coordinated fashion to a range of data-intensive petabyte-scale application domains. This integrating architecture is called data grid [249]. Facing this challenge, some researches have defined the requirements that a data grid must satisfy and the components and APIs that will be required in its implementation.

The paper [249] defines four principles for a data grid architecture.

- Mechanism neutrality. The data grid architecture is designed to be as independent as possible of the low-level mechanisms used to store data, store metadata, transfer data, and so forth.

- Policy neutrality. Within the data grid architecture, data movement and replica cataloging are provided as basic operations, but replication policies are implemented via higher-level procedures. Although default policies are provided, users can easily substitute these policies with application-specific code.

- Compatibility with grid infrastructure. The data grid tools should be compatible with lower-level grid mechanisms. This approach also simplifies the implementation of strategies that integrate, for example, storage and computation.

- Uniformity of information infrastructure. This means that we use the same data model and interface to access the data grid's metadata, replica, and instance catalogs as are used in the underlying grid information infrastructure.

These four principles led us to develop a two layer architecture: core services and high level components. In the core services layer, we have Storage Systems and the Grid Storage API, Metadata Service and other basic services (e.g. authorization and authentication service, resource information service). The replica manager service and the replica selection service are two components in the high level layer. The replica manager service is needed to create (or delete) copies of file instances, or replicas, within specified storage systems. The replica selection service is the process of choosing a replica that will provide an application with data access characteristics.

6.7 Fault tolerant

With the development of grid technology, more and more resources and users join into the community of grid computing. Increasing the number of components in a distributed system means increasing the probability that some of these components will be subject to failure during the execution of a distributed algorithm. To avoid the necessity of restarting an algorithm each time a failure occurs, algorithms should be designed so as to deal properly with such failures [283].

6.7.1 Fault-tolerant algorithms

The distributed system has the *partial failure* property. Because of the dispersion of processing resources in the distributed system, no matter what kind of failure occurs, it usually affects only a part of the entire system. Therefore it is possible that the tasks of failing processes can be taken over by the remaining components, leading to a graceful degradation rather than an overall malfunctioning.

There are two main types of fault-tolerant algorithms, namely robust algorithm and stabilizing algorithm.

Robust algorithm: Robust algorithms are designed to guarantee the correct behavior of the processes which function correctly in spite of faults occurring in other processes during their execution. These algorithms are based on strategies such as the vote or the replication, which maintains the correct behavior of a failure process by other backup processes. These algorithms will never be blocked by the failure of processes, because of the strategies of the vote and the replication. Usually, a robust algorithm is used to deal with permanent faults.

Stabilizing algorithms: Stabilizing algorithms permit the failure of correct processes. Correct processes might be affected by failure, but the algorithm will eventually repair the failure after certain times. The system with these algorithms can be started in any state (possibly faulty), and stabilizing algorithms should finally resume the correct behavior.

In order to determine how a correct process can be affected by the failure, the failure model must be studied. We suppose that only the processes can fail, and the communications between two processes are reliable.

- Initially dead processes. A subset of the processors never ever start.

- Crash model. A process functions properly according to its local algorithm until a certain point where it stops indefinitely.

- Byzantine behavior. The process may execute local algorithms, but the execution sequence of each algorithm is arbitrary. The process may send and receive arbitrary messages.

In the failure model, a *dead process* is the special case of the *crash model*, when the crashed process crashes before it starts executing. Similarly, when the Byzantine process crashes, and then stays in that state, the *byzantine behavior* transforms into the *crash model*.

6.7.2   Fault-tolerant techniques

In order to ensure the performance of the entire system, some techniques must be achieved to recover or replace the process failed. Several fault-tolerant techniques are intensively studied.

Stable memory: With this technique, processes save some recovery points regularly. When a process fails, it is recovered from the last recovery point. This technique is simple to implement, but the problem of cohering globally the entire distributed system is delicate. There are three strategies of saving the recovery points:

1. The process saves recovery points in an asynchronous way, without the dependence of the other processes. In the case of failure, we need a set of recovery points which represent a coherent global state of the system to restart the computing.

2. The recording of the recovery points is preset in order to represent correspondingly a coherent global state. There should be many messages exchanged between processes.

3. Dynamic coordination between the actions of recording of recovery points.

Replication processes: The replication is a technique achieving fault tolerance in a distributed system. The replication is considered an effective means to increase the reliability of a distributed system. Moreover, the replication can improve the performance of the system by using the backup servers. However, the problem of cohering these backup servers constrains these advantages. In other words it is a question of guaranteeing the coherence of servers within acceptable times. Strategies of replication aim at guaranteeing a strong coherence (strong consistency) between the backups of a distributed resource. Formally these strategies ensure that the state of each backup is identical. The passive replication and the active replication are two strategies of replication.

1. The passive replication distinguishes two behaviors from a distributed component: the primary copy and backups. The primary

copy is the only one to carry out all the treatments. The backups supervise passively the primary copy. In the case of failure of the primary copy, a backup becomes the new primary copy.

2. The active replication treats equally each backup of a distributed component. All the backups receive the same sequence, and all the backups process the request independently. After the treatment of the request, the backups send the result to the client autonomously.

3. The semi-active replication is located between the active replication and the passive replication. Contrary to the passive replication, the backups are not passive. All the backups receive the same sequence, and all the backups process the request independently. But after the treatment of the request, the primary copy is the only one to send the result.

6.7.3 Grid fault tolerance

The grid can share numerous distributed resources and provide lots of services for the use of science researchers or e-business. Such services may require several or more days of computation, and execute in a heterogeneous environment. It is therefore necessary to investigate the application of fault-tolerant techniques for grid computing.

Hwang, S. and Kesselman, C. [268] present a failure detection service (FDS) and a flexible failure handling framework (Grid-WFS) as a fault tolerance mechanism on the grid. The FDS enables the detection of both task crashes and user-defined exceptions. By using a notification mechanism which is based on the interpretation of notification messages being delivered from the underlying grid resources, a generic failure detection service can detect failures without requiring any modification to both the grid protocol and the local policy of each grid node. The Grid-WFS built on top of FDS allows users to achieve failure recovery in a variety of ways depending on the requirements and constraints of their applications. Central to the framework is flexibility in handling failures. This section describes how to achieve flexibility by the use of workflow structure as a high-level recovery policy specification, which enables support for multiple failure recovery techniques, the separation of failure handling strategies from the application code, and user-defined exception handlings.

Phoenix [269] is a transparent middleware-level fault-tolerance layer that transparently makes data-intensive grid applications fault tolerant. It detects failures early, classifies failures into transient and permanent, and appropriately handles the transient failures. It also handles information-loss problems associated with building error handling in lower layers by persistently logging failures to the grid knowledge base and allowing sophisticated applications to use this information to tune it. Phoenix is applied to a prototype of the NCSA image processing pipeline and it considerably improves the failure handling and reports on the insights gained in the process.

6.8   Concluding remarks

This chapter first introduces some scheduling algorithms and strategies for heterogeneous computing system. There are 11 static heuristics and two types of dynamic heuristics that are described. Then the scheduling problems in a grid environment are discussed. Concurrently, the grid scheduling algorithms, grid scheduling architecture and some meta-scheduler projects are presented. The convergence of Web service and grid makes the grid service more and more utilized in the business domain. Thus the key components for the grid service scheduling are also introduced. As a specific case of applications scheduling, data-intensive applications scheduling is then presented. Finally, fault-tolerant technologies are discussed to deal properly with system failures and to assure the functionality of the entire grid system.

References

[234] mygrid-directly supporting the e-scientists, 2001. Available online at: http://www.mygrid.org.uk (Accessed 30th September, 2007).

[235] G. Aloisio, M. Cafaro, I. Epicoco, and S. Fiore. Analysis of the globus toolkit grid information service. Technical report, HPCC, University of Lecce, Italy, 2002.

[236] G. Aloisio, M. Cafaro, I. Epicoco, S. Fiore, D. Lezzi, M. Mirto, and S. Mocavero. *Computational Science and Its Applications ICCSA 2005*, chapter Resource and Service Discovery in the iGrid Information Service. Lecture Notes in Computer Science. Springer Berlin/Heidelberg, 2005.

[237] R. Armstrong, D. Hensgen, and T. Kidd. The relative performance of various mapping algorithms is independent of sizable variances in run-time predictions. In *Proceedings of the Seventh Heterogeneous Computing Workshop*, page 79. IEEE Computer Society, 1998.

[238] K. Arnold, B. Osullivan, R. W. Scheifler, J. Waldo, A. Wollrath, B. O'Sullivan, and R. Scheifler. *The Jini(TM) Specification (The Jini(TM) Technology Series)*. Addison-Wesley, 1999.

[239] S. Banerjee, S. Basu, S. Garg, S. Garg, S.-J. Lee, P. Mullan, and P. Sharma. Scalable grid service discovery based on uddi. In *MGC'05: Proceedings of the 3rd international workshop on Middleware for grid computing*, pages 1–6, New York, NY, 2005. ACM Press.

[240] O. Beaumont, L. Carter, J. Ferrante, and Y. Robert. Bandwidth-centric allocation of independent task on heterogeneous plataforms. In *Proceedings of the Internetional Parallel and Distributed Processing Symposium*, Fort Lauderdale, FL, 2002.

[241] T. D. Braun, H. J. Siegel, N. Beck, L. Bni, M. Maheswaran, A. I. Reuther, J. P. Robertson, M. D. Theys, B. Yao, D. A. Hensgen, and R. F. Freund. A comparison study of static mapping heuristics for a class of meta-tasks on heterogeneous computing systems. In *Heterogeneous Computing Workshop*, pages 15–29, 1999.

[242] T. D. Braun, H. J. Siegel, N. Beck, L. Bölöni, M. Maheswaran, A. I. Reuther, J. P. Robertson, M. D. Theys, and B. Yao. A taxonomy for describing matching and scheduling heuristics for mixed-machine heterogeneous computing systems. In *1998 IEEE Symposium on Reliable Distributed Systems*, pages 330–335, 1998.

[243] S. Burke, S. Campana, A. D. Peris, F. Donno, P. M. Lorenzo, R. Santinelli, and A. Sciaba. *GLITE 3 User Guide Manuals Series*, January 2007.

[244] R. Buyya, D. Abramson, and J. Giddy. Nimrod/g: An architecture for a resource management and scheduling system in a global computational grid. In *Proceedings of the HPC ASIA'2000*, China, 2000. IEEE CS Press.

[245] R. Buyya, D. Abramson, J. Giddy, and H. Stockinger. Economic models for resource management and scheduling in grid computing. *Concurrency and Computation: Practice and Experience*, 14:1507–1542, 2002.

[246] R. Byrom and all. R-gma: A relational grid information and monitoring system. In *Proceedings of the Cracow'02 Grid Workshop*, 2003.

[247] H. Casanova, A. Legrand, D. Zagorodnov, and F. Berman. Heuristics for scheduling parameter sweep applications in grid environments. In *Proceedings of the 9th Heterogeneous Computing Workshop*, pages 349–363, Cancun, Mexico, 2000. IEEE Computer Society Press.

[248] D. Chakraborty, F. Perich, S. Avancha, and A. Joshi. Dreggie: Semantic service discovery for m-commerce applications. In *Workshop on Reliable and Secure Applications in Mobile Environment, In Conjunction with 20th Symposium on Reliable Distributed Systems (SRDS)*, 2001.

[249] A. Chervenak, I. Foster, C. Kesselman, C. Salisbury, and S. Tuecke. The data grid: Towards an architecture for the distributed management and analysis of large scientific datasets. *Network and Computer Applications*, 23:187–200, 2001.

[250] K. Cooper, A. Dasgupta, K. Kennedy, C. Koelbel, A. Mandal, G. Marin, M. Mazina, J. Mellor-Crummey, F. Berman, H. Casanova, A. Chien, H. Dail, X. Liu, A. Olugbile, O. Sievert, H. Xia, L. Johnsson, B. Liu, M. Patel, D. Reed, W. Deng, C. Mendes, Z. Shi, A. YarKhan, and J. Dongarra. New grid scheduling and rescheduling methods in the grads project. In *Proceedings of Parallel and Distributed Processing Symposium, 2004.*, 2004.

[251] K. Czajkowski, S. Fitzgerald, I. Foster, and C. Kesselman. Grid information services for distributed resource sharing. In *Proc. 10th IEEE Symp. On High Performance Distributed Computing, 2001.*, 2001.

[252] EGEE Community. EGEE (enabling grids for E-science). Available online at: http://www.eu-egee.org/ (Accessed 30th September, 2007).

[253] M. M. Eshaghian, editor. *Heterogeneous Computing*. Artech House Publishers, 1996.

[254] M. Faerman, R. W. A. Su, and F. Berman. Adaptive performance prediction for distributed data-intensive applications. In *Proceedings of the ACM/IEEE SC99 Conference on High Performance Networking and Computing*, Portland, OH, 1999. ACM Press.

[255] P. Fibich, L. k Matyska, and H. Rudová. Model of grid scheduling problem. In *Exploring Planning and Scheduling for Web Services, Grid and Autonomic Computing*, pages 17–24. AAAI Press, 2005.

[256] S. Fitzgerald, I. Foster, C. Kesselman, G. von Laszewski, W. Smith, and S. Tuecke. A directory service for configuring high-performance distributed computations. In *Proc. 6th IEEE Symp. on High Performance Distributed Computing*, pages 365–375. IEEE Computer Society Press, 1997.

[257] I. Foster. What is the Grid? A three point checklist. *GridToday*, 1(6), 2002.

[258] I. Foster and C. Kesselman, editors. *The Grid: Blueprint for a New Computing Infrastructure*. Morgan Kaufmann Publishers, San Francisco.

[259] I. Foster and C. Kesselman. Globus: A toolkit-based grid architecture. In I. Foster and C. Kesselman, editors, *The Grid: Blueprint for a New Computing Infrastructure*, pages 259–278. Morgan Kaufmann, San Francisco, 1999.

[260] R. F. Freund and H. J. Siegel. Heterogeneous processing. In *Computer*, volume 26, pages 13–17. IEEE Computer Society Press, 1993.

[261] J. Frey, T. Tannenbaum, M. Livny, I. Foster, and S. Tuecke. Condor-g: A computation management agent for multi-institutional grids. *Cluster Computing*, 5(3):237–246, 2002.

[262] Globus Team. Information services (mds): Key concepts. Available online at: `http://www.globus.org/toolkit/docs/4.0/info/key-index.html` (Accessed 30th September, 2007).

[263] GridWay Team. Metascheduling technologies for the grid. Available online at: `http://www.gridway.org/about/visionaim.php` (Accessed 30th September, 2007).

[264] X. He, X. Sun, and G. Laszewski. A qos guided min-min heuristic for grid task scheduling. *Computer Science and Technology*, 18(4):442–451, 2003.

[265] X. He, X.-H. Sun, and G. V. Laszewski. A qos guided scheduling algorithm for grid computing. In *Proceedings of the International Workshop on Grid and Cooperative Computing*, 2002.

[266] R. Henderson and D. Tweten. Portable batch system: External reference specification, 1996.

[267] W. Hoschek. Peer-to-peer grid databases for web service discovery. *Concurrency: Pract. Exper.*, pages 1–7, 2002.

[268] S. Hwang and C. Kesselman. A flexible framework for fault tolerance in the grid. *Journal of Grid Computing*, pages 251–272, 2004.

[269] G. Kola, T. Kosar, and M. Livny. Phoenix: Making data-intensive grid applications fault-tolerant. In *Proceedings of 5th IEEE/ACM International Workshop on Grid Computing*, 2004.

[270] S. Lee, W. Seo, D. Kang, K. Kim, and J. Y. Lee. A framework for supporting bottom-up ontology evolution for discovery and description of grid services. *Expert Systems with Applications*, 32:376–385, 2007.

[271] M. Litzkow, M. Livny, and M. Mutka. Condor-a hunter of idle workstations. In *8th Intl Conference on Distributed Computing Systems*, pages 104–111, 1988.

[272] S. A. Ludwig and P. van Santen. A grid service discovery matchmaker based on ontology description. In *EuroWeb 2002*. British Computer Society, 2002.

[273] M. Maheswaran, S. Ali, H. J. Siegel, D. Hensgen, and R. F. Freund. Dynamic mapping of a class of independent tasks onto heterogeneous computing systems. *Journal of Parallel and Distributed Computing*, 59(2):107–131, 1999.

[274] S. Miles, J. Papay, V. Dialani, M. Luck, K. Decker, T. Payne, and L. Moreau. Personalised grid service discovery. In *Performance Engineering. 19th Annual UK Performance Engineering Workshop*, pages 131–140, 2003.

[275] S. Miles, J. Papay, T. Payne, K. Decker, and L. Moreau. Towards a protocol for the attachment of semantic descriptions to grid services. In *The Second European across Grids Conference*, Nicosia, Cyprus, 2004.

[276] J. Pathak, N. Koul, D. Caragea, and V. G. Honavar. A framework for semantic web services discovery. In *Proceedings of the 7th annual ACM international workshop on Web information and data management*, pages 45–50. ACM Press, 2005.

[277] K. Ranganathan and I. Foster. Decoupling computation and data scheduling in distributed data-intensive applications. In *Proceedings of 11th IEEE International Symposium on High Performance Distributed Computing, 2002. HPDC-11 2002.*, pages 352–358, 2002.

[278] E. Santos-Neto, W. Cirne, F. Brasileiro, and A. Lima. Exploiting replication and data reuse to efficiently schedule data-intensive applications on grids. In *Proceedings of 10th Job Scheduling Strategies for Parallel Processing*, 2004.

[279] H. J. Siegel and S. Ali. Techniques for mapping tasks to machines in heterogeneous computing systems. *Journal of Systems Architecture*, 46:627–639, 2000.

[280] S. Srinivasan, R. Kettimuthu, V. Subramani, and P. Sadayappan. Selective reservation strategies for backfill job scheduling. In *Workshop on Job Scheduling Strategies for Parallel Processing*, volume 2537, pages 55–71. Springer Lecture Notes in Computer Science, 2002.

[281] V. Subramani, R. Kettimuthu, S. Srinivasan, and S. Sadayappan. Distributed job scheduling on computational grids using multiple simultaneous requests. In *Proceedings of High Performance Distributed Computing, 2002 (HPDC-11 2002)*, pages 359–366, 2002.

[282] P. Tang, P. C. Yew, and C. Zhu. Impact of self-scheduling order on performance on multiprocessor systems. In *ICS'88: Proceedings of the 2nd international conference on Supercomputing*, pages 593–603. ACM Press, 1988.

[283] G. Tel. *Introduction To Distributed Algorithms*. Cambridge University Press, 2 edition, January 2004.

[284] B. Tierney, R. Aydt, D. Gunter, W. Smith, V. Taylor, R. Wolski, and M. Swany. A grid monitoring architecture. Technical Report GWD-Perf-16-1, GGF, 2001.

[285] S. C. Timm. Grid service information discovery. In *Joint Workshop: EGEE/OSG/NorduGrid*, 2006.

[286] N. Tonellotto, R. Yahyapour, and P. Wieder. A proposal for a generic grid scheduling architecture. Technical Report TR-0015, Institute on Resource Management and Scheduling, CoreGRID - Network of Excellence, January 2006.

[287] University of Southampton. Grid registry with metadata oriented interface: Robustness, efficiency, security. Available online at: `http://twiki.grimoires.org/bin/view/Grimoires/` (Accessed 30th September, 2007).

[288] L. Wang, H. J. Siegel, V. R. Roychowdhury, and A. A. Maciejewski. Task matching and scheduling in heterogeneous computing environments using a genetic-algorithm-based approach. *Journal of Parallel and Distributed Computing*, 47(1):8–22, 1997.

[289] J. Yu, S. Venugopal, and R. Buyya. A market-oriented grid directory service for publication and discovery of grid service providers and their services. *J. Supercomput.*, 36(1):17–31, 2006.

[290] S. Zhou. Lsf: Load sharing in large-scale heterogeneous distributed systems. In *Workshop on Cluster Computing*, 1992.

Chapter 7

Workflow design and portal

7.1 Overview

The Workflow Management Coalition *WfMC* defines workflow as "The automation of a business process, in whole or part, during which documents, information or tasks are passed from one participant to another for action, according to a set of procedural rules". In the context of grid computing, a grid workflow is defined as a workflow within a grid computing environment. The grid world has evolved from simple toolkits to authenticate users on remote supercomputers, to a service-based architecture intended for activities ranging from supporting large, distributed virtual organizations for e-science to autonomic and on-demand computing for commercial enterprises. According to the definition in the article [300], a grid workflow is "The automation of the processes, which involves the orchestration of a set of Grid services, agents and actors that must be combined together to solve a problem or to define a new service".

Two approaches are proposed to create a grid workflow: manual and automatic. For the manual approach, the workflow management system (e.g. *Triana* and *Kepler*) normally provides a user graphic interface or a user portal to compose application workflows. Users can compose workflows by dragging programming components from toolboxes, and dropping them onto a workspace to create workflows. The created workflow is described by a workflow specification language, such as *WSFL* or *BPEL*. In this approach, the template technology can used to facilitate the workflow composition. Lots of workflow templates are predefined and saved in a database. Users can query the database to find a template that satisfies their needed functions. Then users modify this template to compose their own workflow which reduces the composition time. The other approach is automatic. Given a description of the desired applications, a valid workflow can be created from individual application components that provide full-fledged automatic programming capabilities. In order to achieve automatic workflow generation, application components must be clearly encapsulated and described with semantic descriptions. This approach is fully presented in Chapter 8, *Semantic web*.

Scheduling is an important factor for the efficient execution of computational workflows on grid environments. Three major categories of workflow scheduling architecture are centralized, hierarchical and decentralized scheduling schemes. Two different scheduling algorithms are adopted by the existing workflow systems. Static approaches are used to make global decisions in favor of entire workflow performance, in addition dynamic approaches make decisions for each individual job only when it becomes ready to execute.

At the end of this chapter, the rescheduling mechanism is discussed to achieve fault tolerance and improve the performance of scheduling. In order to hide low-level grid access mechanisms and to make even non-grid-expert users capable of defining and executing workflow applications, some portal technologies are also presented.

7.2   Management systems

Workflows can be distinguished by the method mathematics which describes a workflow. There are three principal representations to present a workflow [300].

Linear Workflow: It is the most basic and most common workflow. A linear workflow can be considered as a sequence of tasks which must be performed in a specified linear order. The task's "output" data can be the "input" of the next task in the workflow and the most common workflow programming tool is a simple script written in Python, Perl, or even Matlab.

Acyclic Graph Workflow: Workflows can be described by an acyclic graph, where nodes of the graph represent a task to be performed and edges represent dependencies between tasks. In an acyclic graph workflow, tasks are partly sequential and concurrent. Some tasks depend upon the completion of several other tasks which may be executed simultaneously. An additional framework is needed to describe such a workflow with a scripting language.

Cyclic Graphs Workflow: Many grid workflow tools correspond to this model where the cycles represent some loop or iteration of control mechanisms. In this case the nodes in the graph are either services or some form of software component instances or represent more abstract control objects and the graph edges represent messages or data streams or pipes that channel work or information between services and components.

Many workflow management systems support the "cyclic graph model" and represent a workflow as "Cyclic Graphs Workflow". Compositional tools based on graphical layout system allow users to move "components", which represent tasks or services, from a palette to an assembly panel. Using typed input and output ports, the programmer connects together the graph and then executes it.

7.2.1 The Triana system

Triana [297] is a graphical Problem Solving Environment (PSE), providing a user portal to enable the composition of scientific applications. Users can compose applications by dragging programming components from toolboxes, and dropping them onto a scratch pad, or workspace. Triana can be used as a grid computing environment and can dynamically discover and choreograph distributed resources, such as Web services, to greater extend its range of functionality. Triana has a highly decoupled modularized architecture and each component can be used individually or collectively by both applications or end-users.

An independent virtual layer, called the GAP, abstracts the underlying middleware or transport bindings from the Triana programmer. GAP allows the advertisement, discovery and communication of Web and P2P Triana Services by the GAP interface.

7.2.2 Condor DAGMan

DAGMan (Directed Acyclic Graph Manager) [298] is a meta-scheduler for Condor. A directed acyclic graph (DAG) can be used to represent a set of programs where the input, output, or execution of one or more programs is dependent on one or more other programs. The programs are nodes in the graph, and the edges identify the dependencies. Condor finds machines for the execution of programs, but it does not schedule programs (jobs) based on dependencies. The Directed Acyclic Graph Manager (DAGMan) is a meta-scheduler for Condor jobs. It manages dependencies between jobs at a higher level than the Condor Scheduler

DAGMan submits jobs to Condor in an order represented by a DAG and processes the results. An input file defined prior to submission describes the DAG, and a Condor submit description file for each program in the DAG is used by Condor. Each node (program) in the DAG needs its own Condor submit description file. DAGMan is responsible for scheduling, recovery, and reporting for the set of programs submitted to Condor.

DAGMan can help with the resubmission of uncompleted portions of a DAG when one or more nodes resulted in failure. If any node in the DAG fails, the remainder of the DAG is continued until no more forward progress can be made based on the DAG's dependencies.

7.2.3 Scientific Workflow management and the Kepler system

Kepler [293] is based on the Ptolemy II system for heterogeneous, concurrent modeling and design. Ptolemy II was developed by the members of the Ptolemy project at UC Berkeley, which provides a mature platform for building and executing workflows, and supports multiple models of computation. With Kepler's intuitive GUI, Kepler can be used by workflow engineers and end users to design, model, execute, and reuse scientific workflows.

Kepler currently provides the following features:

Distributed Execution (Web and Grid services): Kepler provides an interface to seamlessly plug in and execute any WSDL-defined web service. Thus it allows to utilize computational resources distributed in a grid environment.

Database Access and Querying: An interface which can be used to interact with databases is provided.

Other Execution Environments: Kepler supports foreign language interfaces via the Java Native Interface (JNI). It gives the user flexibility to reuse existing analysis components and to target appropriate computational tools.

7.2.4   Taverna in life science applications

Taverna [310] was developed by the $^{my}Grid$ project: a UK e-Science pilot project building middleware to support exploratory, data-intensive, in *silico* experiments in molecular biology. The Taverna workflow workbench environment enables the scientific user to create and run workflows written in the Simplified conceptual workflow language (Scufl). Taverna provides a three-tiered data model for describing resources and their inter-operation: the application data flow layer, the execution flow layer, and the processor invocation layer.

The Taverna workbench provides the main user interface to enable the construction and editing of Scufl workflows, and loading and saving these in an XML serialization (known as XScufl). In order to perform the discovery of distributed services, Taverna supports several service directories and mechanisms, such as UDDI, GRIMOIRES, URL submission, workflow introspection, processor-specific mechanisms and scavenging. Thus it can perform both the syntactic and semantic service discovery. Once the appropriate service components have been located, the user requires an interface allowing them to manually compose these services into a workflow.

7.2.5   Karajan

Karajan [304] is a workflow system which provides a workflow specification language and an execution engine, being developed within the Java CoG Kit. The execution engine is based on the *Globus Toolkit*, while the language is a common purpose XML-based language which can express complicated workflows including conditional control flows and loops.

The execution model includes execution elements of the running tasks, and the events generated during the workflow execution. Execution elements can be in different states depending on the current status of the execution. Events can be generated by the elements or the environment, and cause different actions. Dynamic scheduling of the elements is done in a simple way (list of available resources). Basic checkpointing has also been implemented on the workflow level (the workflow can be checkpointed, when all its elements are in consistent states).

7.2.6   Workflow management in GrADS

The Grid Application Development Software (GrADS) project [314] attempts to provide programming tools and execution environments for ordinary scientific users to develop, execute, and tune applications on the grid. Being a collaboration between several American universities, GrADS supports application development either by assembling domain-specific components from a high-level toolkit or by creating a module by relatively low-level (e.g., MPI) code.

New grid scheduling and rescheduling methods [305] are introduced in GrADS. The scheduler obtains resource information by using information services and locates necessary software on the scheduled node by querying GrADS Information Service (GIS). The scheduler also uses performance models to estimate the performance of the application. With this information, The workflow scheduler ranks each qualified resource for each application component. Lower rank values indicate a better match for the component. After ranking, a performance matrix is constructed and used by the scheduling heuristics to obtain a mapping of components onto resources. Three heuristics have been applied in GrADS; those are Min-Min, Max-Min, and Sufferage heuristics.

The framework of application development has two key concepts. First, applications are encapsulated as configurable object programs (COPs), which include code for the application (e.g. an MPI program), a mapper that determines how to map an application's tasks to a set of resources, and an executable performance model that estimates the application's performance on a set of resources. Second, the system relies upon performance contracts that specify the expected performance of modules as a function of available resources.

GrADS utilizes Autopilot to monitor performance of the agreement between the application demands and resource capabilities. Once the contract is violated, a simple stop/migrate/restart approach and a process-swapping approach are applied to rescheduling grid applications, improving the performance of the system.

7.2.7   Petri net model

A Petri net consists of places, transitions, and directed arcs. Arcs run between places and transitions and connections between two nodes of the same type are not allowed. Places may contain any number of tokens. Transitions act on input tokens by a process known as firing. A transition is enabled if it can fire. When a transition fires, it consumes the tokens from its input places, performs some processing task, and places a specified number of tokens into each of its output places.

Petri nets allow for the design of complex workflows using advanced routing constructs. In the same time, powerful analysis techniques can be used to verify the correctness of a workflow process definition. Unfortunately, most workflow management systems are not based on Petri nets. There are just a few products which use Petri nets as a design language [292].

- Grid-Flow [301] is an infrastructure which supports the design and prototype implementation of a scientific workflow and assists researchers in specifying scientific experiments using a Petri Net-based interface. The Grid-Flow infrastructure is designed as a Service Oriented Architecture (SOA) with multi-layer component models. In order to create a workflow, Grid-Flow provides a new, light-weight, programmable grid workflow language, Grid Flow Description Language (GFDL). A Petri Net-based user interface helps the user design the workflow process with a Petri Net model and a program integration component of the Grid-Flow system is presented to integrate all possible programs into the system.

- Andreas Hoheisel [302] describes an infrastructure developed in the Fraunhofer Resource Grid (FhRG). The FhRG workflow is built on the more expressive formalism of Petri nets. Dynamic workflow graph refinement is introduced as a powerful technique to transform abstract workflow graphs into the concrete ones needed for execution and to automatically add fault tolerance to complex workflows.

  The Fraunhofer Resource Grid is a grid initiative funded by the German federal ministry of education and research. The objective of this project is to develop and to implement a stable and robust grid infrastructure and to provide an easy-to-use interface for controlling distributed applications and services in the grid environment.

7.3   Workflow specification languages

The convergence of grid technologies and Web services promote the emergence of grid services which provide standard mechanisms to solve problems of description, discovery and communication of distributed resources. However, grid services can realize their full potential only if there is a mechanism to dynamically compose new services with existing ones. Thus the workflow languages are needed to describe the various interactions between the services and to define the protocols and the processes of the new services compositions [308].

7.3.1 Web Services Flow Language (WSFL)

WSFL [312], which is proposed by IBM, is an XML language for the description of Web services compositions. WSFL describes the composition of Web services using a flow model and a global model. The flow model defines a series of operations of the composite Web service, and specifies the order in which these operations execute. The global model specifies the interaction pattern of a collection of Web Services and defines how the composite Web service is mapped into the operations of the individual Web services. Using a locator element, the following services can be identified by WSFL:

- services which provide its WSDL definition

- the service implementation is local

- services which have registered in UDDI (Universal Description, Discovery, and Integration)

- the service provider is referenced in a message generated by an activity within the flow

- The service provider is not restricted in any way by the flow model (any)

7.3.2 Grid services flow languages

The grid services flow language (GSFL) [308] is an XML-based language that allows to create grid services workflow in the OGSA framework. It has been defined using XML schemas. GSFL has the following important features.

- Service Providers, which are the list of services taking part in the workflow. Service providers can be located using the locator element, which allows looking up service providers in a number of ways. Services can be located statically and they can also be started up using factories.

- Activity Model, which describes the list of important activities in the workflow.

- Composition Model, which describes the interactions between the individual services. It describes the control and data flow between various operations of the services, and also the direct communication between them in a peer-to-peer fashion.

- Lifecycle Model, which describes the lifecycle for the various activities and the services which are part of the workflow.

7.3.3 XLANG: Web services for business process design

XLANG [308], a proposal by Microsoft Corporation, is a language that is used to model business processes as autonomous agents. XLANG is an XML business process language which provides a way to orchestrate applications and XML Web services into larger-scale, federated applications by enabling developers to aggregate even the largest applications as components in a long-lived business process. An XLANG service description is a WSDL service description with an extension element that describes the behavior of the service as a part of a business process. XLANG service behavior may also rely on simple WSDL services as providers of basic functionality for the implementation of the business process. However, in August of 2002, Microsoft stated that XLang would be superceded by BPEL4WS .

7.3.4 Business Process Execution Language for Web Services (BPEL4WS)

During the summer of 2002, IBM, Microsoft and BEA released a new workflow specification language named BPEL4WS [295]. BPEL4WS represents the merger of two other workflow specification languages, IBM's Web Services Flow Language (WSFL) and Microsoft's XLANG.

BPEL4WS [294] defines a model and a grammar for describing the behavior of a business process and interactions between the process and its partners. Through Web service interfaces, the partners interact with each other and the structure of the relationship at the interface level is encapsulated in what we call a partner link. The coordination of service interactions, states and logics is defined by BPEL4WS to achieve a business goal. BPEL4WS also introduces systematic mechanisms for dealing with business exceptions and processing faults. Finally, a mechanism is introduced to define the treatment of individual or composite activities within a process in cases where exceptions occur or a partner requests reversal.

BPEL4WS is based on several XML specifications: WSDL 1.1, XML Schema 1.0, and XPath1.0. WSDL messages and XML schema type definitions are the data model used by BPEL4WS processes. XPath provides support for data manipulation. All external resources and partners are represented as WSDL services. BPEL4WS provides extensibility to accommodate future versions of these standards, specifically the XPath and related standards used in XML computation.

7.3.5 DAML-S

DAML-S [291] is a DAML+OIL ontology for describing the properties and capabilities of Web Services. DAML+OIL is a semantic Web markup language which enables the creation of arbitrary domain ontologies that support the unambiguous description of Web content. With DAML-S, Web services can be computer-interpretable and the following tasks are achieved:

Discovery: locate a Web service through a registry service

Invocation: execute an identified service by an agent or other service

Inter-operation: add semantic description into the Web service to break down inter-operability barriers

Composition: through automatic selection, compose and inter-operate existing services

Verification: verify service properties

Execution Monitoring: track the execution of complex or composite tasks

In DAML-S, a workflow is a set of Web services that are related to one another via control constructs. These control constructs are block-structured and therefore lack the ordering flexibility provided by BPEL4WS links. Thus a combination of BPEL4WS and DAML-S to create a composite and semantic process is suggested. BPEL4WS exposes a single WSDL interface for the composite process it contains and could therefore be marked-up in DAML-S as an atomic process. This results in the composite process itself, rather than its internal processing being described in DAML-S.

7.4 Scheduling and rescheduling

Scheduling is an important factor for the efficient execution of computational workflows on grid environments. In general, a scientific workflow application can be represented as a direct acyclic graph (DAG), where the node is the individual job and edge represents the inter-job dependence. Both nodes and edges are weighed for computation cost and communication cost respectively. The makespan, which is the total time needed to finish the entire workflow, is used to measure the performance of workflow applications [315].

7.4.1 Scheduling architecture

Based on the workflow reference model [303] proposed by the Workflow Management Coalition (WfMC), Figure 7.1 shows the architecture and functionalities supported by various components of the grid workflow system.

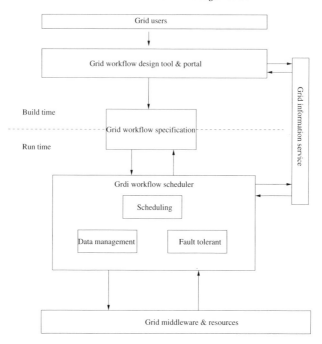

FIGURE 7.1: Grid workflow system architecture.

At the highest level, the grid workflow system may be characterized in three functional areas:

- the Build-time functions define and model the workflow process and its constituent activities. During this phase, a business process is translated from the real world into a formal, computer processable definition by the use of modelling and system definition techniques.

- the Run-time control functions are concerned with managing the workflow processes in an operational environment and mapping each subtask of the workflow into resources. The various activities must be handled such as the data management and fault tolerance.

- the Run-time interactions with human users and IT application tools for processing the various activity steps. Interaction with the process control software is necessary to transfer control between activities, to ascertain the operational status of processes, to invoke application tools and pass the appropriate data, etc.

The architecture of the scheduling infrastructure is very important for scalability, autonomy, quality and performance of the system. Three major categories of workflow scheduling architecture are centralized, hierarchical and decentralized scheduling schemes [314].

In a centralized workflow scheduling environment, one central workflow scheduler makes scheduling decisions for all subtasks in the workflow. The scheduler has the information about the entire workflow and collects information of all available processing resources. The centralized scheme can produce efficient schedules but it is not scalable with respect to the number of tasks and number of grid resources. It is thus only suitable for a small scale workflow.

Unlike centralized scheduling, both hierarchical and decentralized scheduling allow tasks to be scheduled by multiple schedulers. Therefore, one scheduler only maintains the information related to a sub-workflow. Compared to centralized scheduling, the decentralized scheduling is more scalable since they limit the number of tasks managed by one scheduler. However, the best decision made for a partial workflow may lead to sub-optimal performance for the overall workflow execution and there are also some conflict problems. For hierarchical scheduling, there is a central manager and multiple lower-level sub-workflow schedulers. This central manager is responsible for controlling the workflow execution and assigning the sub-workflows to the low-level schedulers. The scheme has more advantages such as scalability and scheduling policies' independence. However, the failure of the central manager will result in entire system failure.

7.4.2 Scheduling algorithms

The existing workflow systems are going into two different scheduling algorithms. Some systems use static approaches by which the scheduler makes the global decisions in favor of entire workflow performance relying on knowledge of the entire DAG and execution environment. Others depend on dynamic approaches by which the scheduler makes decisions for each individual job only when it becomes ready to execute. This type of decision is also referred to as local just-in-time decision [315].

A large number of static scheduling heuristics has been presented in the literature such as min-min, max-min, sufferage and HEFT (Heterogeneous Earliest Finish Time). However, in a grid environment static strategies may perform poorly because of the grid dynamics: a resource can join and leave at any time; individual resource capability varies over time because of internal or external factors. To overcome these limitations of static approach, an HEFT-based adaptive rescheduling algorithm is presented in [315]. With this algorithm, the executor will notify the planner of any run-time event which interests the planner, for example, resource unavailability or discovery of a new resource. In turn, the planner responds to the event by means of evaluating the event and rescheduling the remaining jobs in the workflow if necessary. Planning is now an iterative (event-driven) activity instead of one-time task.

In the GrADS infrastructure, workflow scheduling is based on static heuristic scheduling strategies that use application component performance models [309]. In order to map workflows onto resources, a two-stage approach is presented. In the first stage, a specific rank value is assigned for each resource on which the component can be mapped. Rank values reflect the expected performance of a particular component on a particular resource. In the second stage, a performance matrix is built for each component. Then certain known heuristics are used to obtain a mapping of components to resources. To estimate the execution cost of an application on arbitrary grid configurations, the behavior of applications is analyzed by modeling its characteristics in isolation of any architectural details. Then the application's execution cost is estimated on a target platform.

While sizable work supports the claim that the static scheduling performs better for workflow applications than the dynamic one, the dynamic approach is needed in environments where high inaccuracies are observed. Yu, J. and Buyya, R. [313] propose a workflow enactment engine (WFEE) with a just-in-time scheduling system using tuple spaces. It allows the decision of resources allocation to be made dynamically at the time of the execution of tasks in the workflow. A decentralized event-driven scheduling architecture provides a flexible and loosely-coupled control. In this system, every task has its own scheduler called task manager (TM) which implements a scheduling algorithm and handles the processing of the task, including resource selection, resource negotiation, task dispatcher and failure processing. The lifetimes of TMs and the whole workflow execution are controlled by a workflow coordinator (WCO). Dedicated TMs are created by WCO for each task. Each TM has its own monitor which is responsible for monitoring the health of the task execution on the remote node. Every TM maintains a resource group which is a set of resources that provide services required for the execution of an assigned task. TMs and WCO communicate through an event service server (ESS).

7.4.3 Decision making

There is no single best solution for mapping workflows onto resources for all workflow applications, since the applications can have very different characteristics. In general, decisions about mapping tasks in a workflow onto resources can be based on the information of the current task or of the entire workflow and can be of two types, namely local decision and global decision [314].

Local decision-based scheduling only takes one task or sub-workflow into account, so it may produce the best schedule for the current task or sub-workflow but could also reduce the entire workflow performance. We can assume that there is a data-intensive application where the overall run-time is driven by data transfer costs. In the case where the output of a task is very large, the initial selection may be found to be a poor choice if latency between the nodes is very high, because the selection of a resource for a task is based only on a local decision without consideration of data transfer between other resources. This would lead to higher data transfer costs for child tasks and hence the entire workflow.

Scheduling workflow tasks using global decision improves the performance of the entire workflow. It is believed that global decision-based scheduling can provide a better overall result. However, it may take much more time in scheduling decision making. Thus, the overhead produced by global scheduling could reduce the overall benefit and may even exceed the benefits it will produce. Therefore, the choice of decision making for workflow scheduling should not be made without considering balance between the overall execution time and scheduling time. However, for some applications such as a data analysis application where the outputs of tasks in the workflow are always smaller than the inputs, using local decision-based scheduling is sufficient.

7.4.4 Scheduling strategies

There are three major strategies of scheduling: performance-driven, market-driven and trust-driven [314].

- Performance-driven strategies try to find a mapping of workflow tasks onto resources that achieves optimal execution performance such as minimized makespan. Most grid workflow scheduling systems fall in this category.

- Market models are employed in market-driven strategies to manage resource allocation for processing workflow tasks. Workflow schedulers act as consumers buying services from the resource providers and pay some notion of electronic currency for executing tasks in the workflow. The tasks in the workflow are dynamically scheduled at run-time depending on resource cost, quality and availability, to achieve the desired level of quality for deadline and budget. Unlike the performance-driven strategy, market-driven schedulers may choose a resource with a later deadline if its usage price is cheaper.

- Trust-driven schedulers select resources based on their trust levels. The trust model of resources is based on attributes such as security policy, accumulated reputation, self-defense capability, attack history, and site vulnerability. By using trust-driven approaches, workflow management systems can reduce the chance of selecting malicious hosts and non-reputable resources. Therefore, overall accuracy and reliability of workflow execution will be increased.

7.4.5 Rescheduling

The rescheduling mechanism is an essential component in the workflow system to achieve fault tolerance and improve the performance of scheduling. Two rescheduling mechanisms are described in [305]: Rescheduling by Stop and Restart and Rescheduling by Processor Swapping.

7.4.5.1 Rescheduling by Stop and Restart

In the stop/restart approach, the application is suspended and migrated only when better resources are found for application execution. When a running application is signaled to migrate, all application processes checkpoint user-specified data and terminate. The rescheduled execution is then launched by restarting the application on the new set of resources, which then read the checkpoints and continue the execution.

A user-level checkpointing library must be employed to provide application migration support. Via this library, the application can checkpoint data, be stopped at a particular execution point, be restarted later on a different processor configuration and be continued from the previous point of execution. The information service system is also needed to evaluate the performance of the resource on which the application is migrated. The rescheduler operates in two modes: migration on request and opportunistic migration. When the monitor detects unacceptable performance loss for an application, it contacts the rescheduler to request application migration. This is called migration on request. Additionally, the rescheduler periodically checks for an application that has recently completed. If it finds one, the rescheduler determines if another application can obtain performance benefits if it is migrated to the newly freed resources. This is called opportunistic rescheduling.

7.4.5.2 Rescheduling by Processor Swapping

The stop/restart approach is very flexible but it can be expensive: each migration event can involve large data transfers. The processor swapping approach provides an alternative that is lightweight and easy to use, but less flexible than the stop/restart migration approach.

To enable swapping, the application is launched with more machines than will actually be used for the computation; some of these machines become part of the computation (the active set) while some do nothing initially (the inactive set). During execution, the monitor periodically checks the performance of the machines and swaps slower machines in the active set with faster machines in the inactive set. This approach requires little application modification and provides an inexpensive fix for many performance problems.

7.5 Portal integration

The subtasks of workflow can be mapped onto different grids in order to provide more parallelism than inside one grid. Moreover high-level graphical interfaces must be supplied to hide low-level grid access mechanisms, making even users not expert in grids capable of defining and executing distributed applications on multi-institutional computing infrastructures. The portal technology provides a suitable solution to overcome these questions. It solves the problems of inter-operability between different grids in the workflow level and it offers the workflow GUI as the interface that enables the

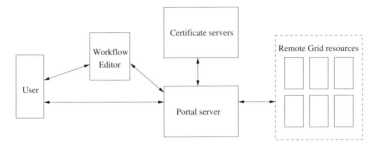

FIGURE 7.2: P-GRADE portal system functions.

development, submission and steering of workflows and the visualization of results.

7.5.1 P-GRADE portal

P-GRADE grid portal [306] is the first grid portal that tries to solve the inter-operability problem at the workflow level with great success. It is a workflow-oriented grid portal with the main goal to support all stages of grid workflow development and execution processes.

The P-GRADE Portal provides the following functions (see Figure 7.2): communicating with the portal server, users can achieve the functions of defining grid environments, managing grid certificates, controlling the execution of workflow applications and visualizing the progress of workflows; Workflow editor can perform the creation and modification of workflow applications [307].

During workflow editing the user has the possibility to select a grid resource for each job, or let a broker choose one. Currently there are two brokers used by the portal: the LCG-2 Broker and GTbroker. The GTbroker interacts with the Globus resources to perform job submission. The static and dynamic information of grid resources are collected by GTbroker to achieve scheduling activities. The LCG-2 broking solution is used to reach LCG-2 based grids. The mission of the LHC Computing Project (LCG) is to build and maintain a data storage and analysis infrastructure for the entire high energy physics community that will use the LHC. The Large Hadron Collider (LHC), currently being built at CERN near Geneva, is the largest scientific instrument on the planet and it begins operations in 2007. With exploiting the broking functions of GTbroker and LCG-2 Broker, users can develop and execute multi-grid workflows in a convenient environment.

The integration of P-GRADE into GEMLCA shows the use of portal in a grid environment [306]. GEMLCA (Grid Execution Management for Legacy Code Applications) represents a general architecture for deploying legacy applications as grid services without re-engineering the code or even requiring access to the source files. GEMLCA adds an additional layer to wrap the legacy application on top of a service-oriented grid middleware, like Globus Toolkit version 4 (GT4). GEMLCA communicates with the client through SOAP-XML messages, gets input parameter values, submits the legacy executable to a local job manager like Condor or PBS (Portable Batch System), and returns the results to the client in SOAP-XML format. GEMLCA provides the capability to convert legacy code into grid services. However, an end-user without specialist computing skills still requires a user-friendly Web interface (portal) to access the GEMLCA functionalities. In order to solve the third problem, GEMLCA is integrated with the P-GRADE grid portal. Following this integration, legacy code services can be included in end-user workflows, running on different GEMLCA grid resources. The workflow manager of the portal contacts the selected GEMLCA resource and passes the actual parameter values of the legacy code to it. Then the GEMLCA resource executes the legacy code with these actual parameter values and delivers the results back to the portal.

7.5.2   Other portal systems

The *Pegasus* [311] Portal provides an HTTP(S)-based interface that can be accessed using a standard web browser. The portal architecture is composed of three layers. The top layer consists of the user machines and web browsers. The second layer consists of the web application server hosting the portal. The server is multi-threaded and can handle multiple user requests at the same time. The third layer consists of the grid components and services used by the portal.

In order to use the Pegasus grid portal the user needs to have a valid grid credential in a MyProxy server. The portal does not provide access to a predetermined set of resources. Instead, the user can specify the resources to be used. From the web browser, the users specify the parameters of the application and Pegasus does the mapping of tasks in the workflow to resources specified in the resource configuration. The submitted workflow may take a long time to complete. The user may logout from the portal and login later to check its status. The portal also allows users to view the status of the workflow (submitted, active, done, failed), the number of tasks already completed, the tasks currently executing, and other information.

The Pegasus grid portal is very useful in scenarios where a virtual organization (VO) wants to provide an easy-to-use application submission interface to its members. It is able to map abstract workflow onto physical resources, thus users are shielded from the complexity of installing and using the various components in order to access the grid resources.

Grid-Flow [296] is a grid workflow management system developed at the University of Warwick. Rather than focusing on workflow specification and the communication protocol, Grid-Flow is more concerned about service-level scheduling and workflow management. The Grid-Flow portal performs two level managements: global grid workflow management and local grid sub-workflow scheduling. The execution and monitoring functionalities are provided at the global grid level, which work on top of an existing agent-based grid resource management system. At each local grid, sub-workflow scheduling and conflict management are processed on top of an existing performance prediction-based task scheduling system. A fuzzy timing technique is applied to address new challenges of workflow management in a cross-domain and highly dynamic grid environment.

7.6 A case study on the use of workflow technologies for scientific analysis

7.6.1 Motivation

The Laser Interferometer Gravitational Wave Observatory (LIGO) [299] is an ambitious effort to detect gravitational waves produced by violent events in the universe, such as the collision of two black holes, or the explosion of a supernova. The experiment records approximately 1 TB of data per day which is analyzed by scientists in a collaboration which spans four continents. These large data sets need to be accessed by various elements of the analysis workflows. In order to transparently execute jobs at remote locations, it is important to have seamless management of jobs and data transfer.

7.6.2 The LIGO data grid infrastructure

To fully leverage the distributed resources in an integrated and seamless way, infrastructure and middleware have been deployed to structure the resources as a grid. The LIGO data grid infrastructure includes the Linux clusters, the networks that interconnect them to each other, grid services running on the LSC Linux clusters, a system for replicating LIGO data to computing centers, grids certificate authority authentication, and a package of client tools and libraries that allow scientists to leverage the LIGO data grid services.

7.6.3 LIGO workflows

Although huge quantities of data must be analyzed over a vast parameter space, analysis does not require interprocess communication. Analysis can be broken down into units that perform specific tasks which are implemented as individual programs, usually written in the C programming language or the Matlab processing environment. Workflows may be parallelized by splitting the full parameter space into smaller blocks, or parallelizing over the time intervals being analyzed. The individual units are chained together to form a

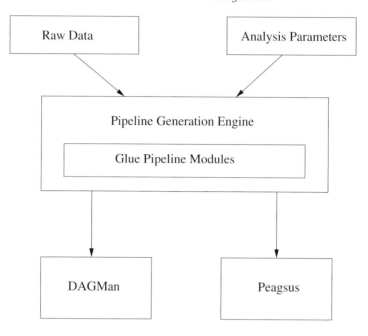

FIGURE 7.3: The creation of pipeline workflow.

data analysis pipeline. The pipeline starts with raw data from the detectors, executes all stages of the analysis and returns the results to the scientist.

Figure 7.3 shows the creation of pipeline workflow. The data analysis pipeline is implemented as a directed acyclic graph (DAG) and several workflow management systems are supported, such as Condor DAGMan and Pegasus. In order to facilitate the workflow design and to abstract the representation of the workflow, Glue was developed especially for scientists to help build workflows. In this way, the components of the workflow are abstracted, and it is straightforward to write pipeline scripts that construct complex workflows. The Glue method of constructing data analysis pipelines has been used in the binary inspiral analysis, the search for gravitational wave bursts from cosmic strings, excess power burst analysis, and in the stochastic gravitational wave background analysis.

7.7   Concluding remarks

Grid workflow is increasingly used to compose complex applications in a grid environment. In this chapter, we first introduce workflow management systems. Then workflow specification languages which are used to describe the operations and dependencies of the workflow components are presented. Next, we discussed the workflow scheduling and rescheduling which are key factors to improve the performance of workflow applications. Then portal projects are introduced and we point out that the portal is an important component to reduce the workflow composition time for the non-expert users. Finally, a use case, LIGO data grid infrastructure, is presented to illustrate the utilization of grid workflow.

References

[291] A. Ankolekar et al. Daml-s: Web service description for the semantic web. In *Proc. 1st Int. Semantic Web Conf. (ISWC 02)*, 2002.

[292] W. Aalst. The application of petri nets to workflow management. *The Journal of Circuits, Systems and Computers*, 8(1):21–66, 1998.

[293] I. Altintas, C. Berkley, E. Jaeger, M. Jones, B. Ludäscher, and S. Mock. Kepler: An extensible system for design and execution of scientific workflows. In *16th Intl. Conf. on Scientific and Statistical Database Management (SSDBM'04)*, 2004.

[294] T. Andrews, F. Curbera, H. Dholakia, Y. Goland, J. Klein, F. Leymann, K. Liu, D. Roller, D. Smith, S. Thatte, I. Trickovic, and S. Weerawarana. Business process execution language for web services (version 1.1), 2003. Available online at: `http://www-106.ibm.com/developerworks/webservices/library/ws-bpel/` (Accessed 30th September, 2007).

[295] P. A. Buhler and J. Vidal. Towards adaptive workflow enactment using multiagent systems. *Information Technology and Management*, 6:61–87, 2005.

[296] J. Cao, S. A. Jarvis, S. Saini, and G. R. Nudd. Gridflow: Workflow management for grid computing. In *Proceedings of the 3st International Symposium on Cluster Computing and the Grid, CCGRID '03*, page 198. IEEE Computer Society, 2003.

[297] D. Churches, G. Gombas, A. Harrison, J. Maassen, C. Robinson, M. Shields, I. Taylor, and I. Wang. Programming scientific and distributed workflow with triana services: Research articles. *Concurr. Comput. : Pract. Exper.*, 18(10):1021–1037, 2006.

[298] Condor Team. Dagman applications. Available online at: `http://www.cs.wisc.edu/condor/manual/v6.4/2_11DAGMan_Applications.html` (Accessed 30th September, 2007).

[299] Duncan A. Brown et al. A case study on the use of workflow technologies for scientific analysis: Gravitational wave data analysis. In *Workflows for e-Science: Scientific Workflows for Grids*. Springer-Verlag, 2006.

[300] G. C. Fox and D. Gannon. Workflow in grid systems. *Concurrency and Computation: Practice & Experience*, 18(10):1009–1019, 2006.

[301] Z. Guan, F. Hernandez, P. Bangalore, J. Gray, A. Skjellum, V. Velusamy, and Y. Liu. Grid-flow: a grid-enabled scientific workflow system with a petri-net-based interface. *Concurrency and Computation: Practice & Experience*, 18:1115–1140, 2006.

[302] A. Hoheisel. User tools and languages for graph-based grid workflows. In *Proceedings of Workflow in Grid Systems Workshop in GGF10*, 2004.

[303] D. Hollingsworth. Workflow management coalition, the workflow reference model, 1994. Document Number TC00-1003, Available online at: `http://www.wfmc.org/standards/docs/tc003v11.pdf` (Accessed 30th September, 2007).

[304] Java CoG Kit Team. Java cog kit karajan workflow reference manual, 2007. Available online at: `http://www.gridworkflow.org/snips/gridworkflow/space/Karajan` (Accessed 30th September, 2007).

[305] K. Cooper et al. New grid scheduling and rescheduling methods in the grads project. In *Proc. IPDPS 2004*, 2004.

[306] P. Kacsuk, T. Kiss, and G. Sipos. Solving the grid interoperability problem by p-grade portal at workflow level. In *Proc. of the 15th IEEE International Symposium on High Performance Distributed Computing (HPDC-15), Paris, France*, 2006.

[307] A. Kertesz, Z. Farkas, P. Kacsuk, and T. Kiss. Multiple broker support by grid portals. In *CoreGRID Workshop on Grid Middleware, Tools and Environments in conjunction with GRIDS@Work, Sophia Antipolis, France*, 2006.

[308] S. Krishnan, P. Wagstrom, and G. von Laszewski. GSFL: A Workflow Framework for Grid Services. Technical report, Argonne National Laboratory, 9700 S. Cass Avenue, Argonne, 1L 60439, U.S.A., 2002.

[309] A. Mandal, K. Kennedy, C. Koelbel, G. Marin, J. Mellor-Crummey, B. Liu, and L. Johnsson. Scheduling strategies for mapping application workflows onto the grid. In *Proceedings of 14th IEEE International Symposium on High Performance Distributed Computing, 2005. HPDC-14*, pages 125–134, 2005.

[310] T. Oinn, M. Greenwood, M. Addis, M. N. Alpdemir, J. Ferris, K. Glover, C. Goble, A. Goderis, D. Hull, D. Marvin, P. Li, P. Lord, M. R. . Pocock, M. Senger, R. Stevens, A. Wipat, and C. Wroe. Taverna: Lessons in creating a workflow environment for the life sciences. *Concurrency and Computation: Practice and Experience*, 18:1067–1100, 2005.

[311] G. Singh, E. Deelman, G. Mehta, K. Vahi, M. Su, B. Berriman, J. Good, J. Jacob, D. Katz, A. Lazzarini, K. Blackburn, and S. Koranda. The pegasus portal: Web based grid computing. In *The 20th Annual ACM Symposium on Applied Computing, Santa Fe, New Mexico*, 2005.

[312] A. Wesley. WSFL in action, Part 1, 2002. Available online at: `http://www-128.ibm.com/developerworks/webservices/library/ws-wsfl1/` (Accessed 30th September, 2007).

[313] J. Yu and R. Buyya. A novel architecture for realizing grid workflow using tuple spaces. In *Proceedings of the Fifth IEEE/ACM International Workshop on Grid Computing (GRID'04)*, pages 119–128. IEEE Computer Society, 2004.

[314] J. Yu and R. Buyya. A taxonomy of workflow management systems for grid computing. Technical Report GRIDS-TR-2005-1, Grid Computing and Distributed Systems Laboratory, University of Melbourne, Australia, 2005.

[315] Z. Yu and W. Shi. An adaptive rescheduling strategy for grid workflow applications. In *Proceedings of the 21st IPDPS 2007, Long Beach, CA*, 2007.

Chapter 8

Semantic web

8.1 Introduction

More than twenty years ago, the World Wide Web was still a blueprint. Nowadays millions of people around the world could not imagine their lives without it. The WWW has quite rapidly evolved into a vast information, communication and transaction space. Unfortunately this growth runs into a problem: lack of a global system which can be used to easily publish and process data in a way standard by anyone. It can sometimes be difficult to find, access, present and maintain the information required by a wide variety of users. One of the main obstacles is that most information on the Web is made for human interpretation and is not evident for agents browsing the Web. The Semantic Web is an effort to improve the current Web by making Web resources "machine-understandable", because the current Web resources do not respect machine-understandable semantics.

8.1.1 Web and semantic web

The World Wide Web (or the "Web") was created around 1990 by the Briton Tim Berners-Lee and the Belgian Robert Cailliau working at CERN in Geneva, Switzerland. The Web is a system of interlinked, hypertext documents that runs over the Internet. With a Web browser, a user views Web pages that may contain text, images, and other multimedia and navigates between them using hyperlinks. Since the creation of the Web, Berners-Lee has played an active role in guiding the development of Web standards (such as the markup languages in which Web pages are composed), and in recent years has advocated his vision of a Semantic Web.

According to Tim Berners-Lee's definition, a Semantic Web is *"an extension of the current web in which information is given well-defined meaning, better enabling computers and people to work in co-operation. It is the idea of having data on the Web defined and linked in a way that it can be used for more effective discovery, automation, integration and reuse across various applications... data can be shared and processed by automated tools as well as by people"*. At its core, the semantic web comprises a set of elements which include Resource Description Framework (RDF), a variety of data interchange formats (e.g RDF/XML, N3, Turtle, N-Triples), and notations such as RDF Schema (RDFS) and the Web Ontology Language (OWL). All of which are intended to formally describe concepts, terms, and relationships within a given knowledge domain.

The Semantic Web is an extension of the current Web. Normally, there are two conceptual differences between the Semantic Web and the Web:

- The Semantic Web is an information space in which the information is expressed in a special machine-targeted language, while the Web is an information space that contains information which aims at human consumption expressed in a wide range of natural languages.

- The Semantic Web is a web of formally and semantically interlinked data, whereas the Web is a set of informally interlinked information.

8.1.2   Ontologies

Thomas R. Gruber [323] provides a definition of ontology as "A specification of a representational vocabulary for a shared domain of discourse – definitions of classes, relations, functions, and other objects". An ontology is a description (like a formal specification of a program) of the concepts and relationships that can exist for an agent or a community of agents.

The term of ontology is borrowed from philosophy, where an ontology is a systematic account of existence. When the knowledge of a domain is represented in a declarative formalism, the set of objects that can be represented is called the universe of discourse. This set of objects, and the describable relationships among them, are reflected in the representational vocabulary with which a knowledge-based program represents knowledge. Thus, in this context, we can describe the ontology of a program by defining a set of representational terms. In such an ontology, definitions associate the names of entities in the universe of discourse (e.g., classes, relations, functions, or other objects) with human-readable text describing what the names mean, and formal axioms that constrain the interpretation and well-formed use of these terms. Formally, an ontology is the statement of a logical theory [323].

An ontology language is a formal language used to encode the ontology. There are a number of such languages for ontologies, both proprietary and standards-based:

- OWL is a language for making ontological statements, developed as a follow-on from RDF and RDFS, as well as earlier ontology language projects including OIL, DAML and DAML+OIL. OWL is intended to be used over the World Wide Web, and all its elements (classes, properties and individuals) are defined as RDF resources, and identified by URIs.

- KIF is a syntax for first-order logic that is based on S-expressions.

- The Cyc project has its own ontology language called CycL, based on first-order predicate calculus with some higher-order extensions.

8.1.2.1   RDF

Resource Description Framework (RDF) is a family of World Wide Web Consortium (W3C) specifications originally designed as a metadata model but which has come to be used as a general method of modeling information, through a variety of syntax formats. In RDF the information maps directly and unambiguously to a decentralized model, for which there are many generic parsers already available. Thus for an RDF application, you know which bits of data are the semantics of the application, and which bits are just syntactic, in turn enabling users to deal with the information with greater efficiency and certainty [331].

The RDF metadata model is based upon triples in RDF terminology. The subject denotes the resource, and the predicate denotes traits or aspects of the resource and expresses a relationship between the subject and the object. For example, one way to represent the notion "The sky has the color blue" in RDF is as a triple of specially formatted strings: a subject denoting "the sky", a predicate denoting "has the color", and an object denoting "blue".

8.1.2.2   OWL

The Web Ontology Language (OWL) is a language for defining and instantiating Web ontologies and is designed for use by applications that need to process the content of information instead of just presenting information to humans. An OWL ontology may include descriptions of classes, along with their related properties and instances and it is developed to augment the facilities for expressing semantics (meaning) provided by XML, RDF, and RDF-S. OWL is based on earlier languages OIL and DAML+OIL, and is now a W3C recommendation.

OWL is seen as a major technology for the future implementation of a Semantic Web. It is playing an important role in an increasing number and range of applications, and is the focus of research into tools, reasoning techniques, formal foundations and language extensions.

OWL is designed to provide a common way to process the semantic content of web information. It may be considered an evolution of these web languages in terms of its ability to represent machine-interpretable semantic content on the web. Since OWL is based on XML, OWL information can be easily exchanged between different types of computers using different operating systems, and application languages. Because the language is intended to be read by computer applications, it is sometimes not considered to be human-readable. OWL is being used to create standards that provide a framework for resource management, enterprise integration, and data sharing on the Web.

An extended version of OWL (sometimes called OWL 1.1, but with no official status) has been proposed which includes increased expressiveness, a simpler data model and serialization, and a collection of well-defined sublanguages each with known computational properties. OWL 1.1 has a well-defined model-theoretic semantics, and is motivated by application requirements. The major builders of the Semantic Web, namely, RACER, FaCT++, Pellet and Cerebra expressed a commitment to support OWL 1.1 in the near future.

8.1.2.3 DAML+OIL

DAML+OIL is a more recent proposal for an ontology representation language that has emerged from work under DARPA's Agent Markup Language (DAML) initiative along with input from leading members of the OIL consortium. DAML+OIL is the basis of the W3C Web Ontology Language OWL, and its predecessor. DAML+OIL is an ontology language which is specifically designed for use on the Web, exploits existing Web standards (XML and RDF) and adds the formal description logic. DAML+OIL is designed on top of the object oriented approach, describing the structure in terms of classes and properties. The ontology of DAML+OIL consists of a set of axioms (e.g., asserting class subsumption/equivalence), and DAML+OIL classes can be names (URIs) or expressions. A variety of constructors are provided for building class expressions. The expressive power of the language is determined by the class (and property) constructors supported, and by the kinds of axiom supported [324].

As a delivery platform for ontologies, DAML+OIL is quite satisfactory and indeed, in the opinions of the authors, is a great improvement over alternative representations such as simple RDF schema or topic maps. However, as an exchange and modeling format, DAML+OIL is lacking the mechanism of ensuring the suitability and intelligibility of knowledge for the users through transformations [320].

8.2 Semantic grid

8.2.1 The grid and the semantic web

The concept of the grid emerged in the early 1990s as a distributed infrastructure for advanced science and engineering. With the development of grid technologies, three generations of the grid can be identified: first generation systems focus on solutions for sharing high performance computing resources, second generation systems introduce middleware to deal with the scale and heterogeneity of large-scale computational power and large volumes of data and third generation systems are adopting a service-oriented approach which can achieve the standardization for the grid and allow a secure and robust infrastructure to be built.

Although grid computing is gaining a lot of attention within the IT industry, grid computing is not yet a standard product on the ICT (Information and Communications Technology) market. Presently there are only a few real production grids and current grid middleware is extremely hard to use for non-specialist users and incomplete. It has provided computational inter-operability, but semantic inter-operability is now required. To overcome present architectural and design limitations which prevent the use and wider deployment of computing and knowledge grids and to enrich its capabilities by including new functionalities required for complex problem solving, the larger uptake of grid-type architectures is needed and the concept from computation grids to knowledge grids must be extended, eventually leading to a "semantic grid".

The integration of semantic technology increases the inter-operability of the Web and grid, and the Semantic grid is the convergence of the Semantic Web and the grid, as shown in Figure 8.1. The Semantic grid refers to an approach to grid computing in which information, computing resources and services are described in standard ways that can be processed by computer. This makes it easier for resources to be discovered and joined up automatically, which helps bring resources together to create virtual organizations. The descriptions constitute metadata and are typically represented using the technologies of the Semantic Web, such as the Resource Description Framework (RDF). The use of Semantic Web and other knowledge technologies in grid applications is sometimes described as the Knowledge Grid. Semantic Grid extends this by also applying these technologies within the grid middleware.

The semantic ability can be applied in two aspects of the Semantic Grid: the discovery of available resources and the data integration [317].

Discovery and reuse of resources The discovery of available resources helps grid users reuse existing resources and technology for their grid requirements, instead of new grids and applications being built for processing new data. Rather than developing a single-use application for a grid environment, the existing grid infrastructure and grid application could be made available to other people. In a Semantic Grid, a user or application must provide a more detailed definition of the capabilities required to find and make use of the grid. The semantic technologies, such as RDF and OWL, will be employed to make it easier to determine what facilities a specific grid can provide.

Data integration The semantic technology can be used to provide links and connectivity between information stored and available within a data grid. Thus it provides an efficient way of storing and retrieving information. In addition, the definition of the data stored and processed by a grid will enable users to string together multiple grids to provide complex calculations.

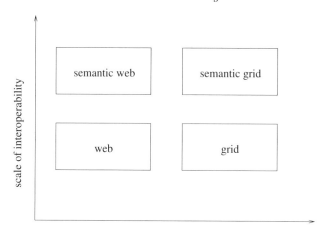

scale of data and computation

FIGURE 8.1: Convergence of the semantic web and the grid technologies.

8.2.2   Current status of the semantic grid

At present Semantic Web technologies such as the Resource Description Framework (RDF) for metadata representation are increasingly being applied to grid computing infrastructures and applications, facilitating interoperability and reuse of services, data and tools. Some research and projects have been done to try to integrate semantic technologies into the grid, in order to achieve automatically the resource discovery and data mining. However it is agreed that the Semantic Grid still has a long way to go to become a reality.

Yolanda Gil and her colleagues describe AI techniques that automatically generate complex, detailed workflows that can execute on a grid [322]. Currently, high-level services such as workflow generation and management systems lack expressive descriptions of grid entities and their relationships. The workflow generation and mapping system which is presented by Yolanda Gil and her colleagues integrates an AI planning system into a grid environment. A user submits an application-level description of the desired data product. Thus for the AI planning system, the goals are the desired data products and the operators are the application components. The AI planning system then receives as input a representation of the current state, a declarative representation of a goal state, and a library of operators that the planner can use to change the state. The goal of the planning system is to search for a valid, partially ordered set of operators that will transform the current state into one that satisfies the goal. The plan returned corresponds to an executable workflow, which includes the assignment of components to specific resources that can be executed to provide the requested data product.

In the industry, scientists normally need a deep understanding of the microscopic structure of components, detailed analytical information about how the various components interact, and the ability to predict how the components will perform under load conditions. To solve these problems, scientists are turning to knowledge management techniques, particularly Semantic Web technologies, to make sense of and assimilate the vast amounts of microstructural, performance, and manufacturing data that they acquire during their research for mining and discovery of novel relations, and for advanced management techniques. Jane Hunter, John Drennan, and Suzanne Little's article [325] exemplifies the need for image analysis and visualization techniques to support scientific and engineering advances. It describes a project that uses Semantic Web technology to develop novel hydrogen-based energy sources.

Chris Wroe and his colleagues describe how they use Semantic Web languages to integrate services and data on grid computing. Their work supports interactive bioinformatics experimentation by pulling information from distributed heterogeneous services and integrating it through expressive semantic representations of their content and the operations they support. They introduce the notion of scientific workflows. These workflows consist of individual processing steps whose results are consumed by others and that overall generate the desired data products of scientific analysis.

Mario Cannataro and Domenico Talia [318] believe the grid is moving from computation and data management to a pervasive, worldwide knowledge management infrastructure. A comprehensive software architecture for the next-generation grid which integrates currently available services and components in Semantic Web, Semantic Grid, P2P, and ubiquitous systems is proposed. Knowledge discovery and data mining, ontology-driven organization of grid-related knowledge, and intelligent data exploration and visualization will be part of future advanced applications of grid-computing infrastructure for science.

8.2.3   Challenges to be overcome

Achieving the full richness of the Semantic Grid vision brings with it many significant research challenges. Some of the technical ones have been highlighted. However others arise from the need to bring together the research communities to achieve the Semantic Grid ambitions. Communities must be grouped together to create the Semantic Grid, which can then be used for flexible collaborations and computations on a global scale for the creation of new scientific results, new business and even new research disciplines.

David De Roure and his colleagues describe ten research challenges in the Semantic Grid which need to be targeted in the future [334] such as Automated Virtual Organization Formation and Management, Service Negotiation and Contracts, Metadata and Annotation and Pervasive Computing, etc. Moreover, we want to highlight some of the technical ones. In Semantic Grid, we use ontologies to express inter-operability between systems. Tools are now appearing that facilitate the construction of a verification of ontologies. The Semantic Web effort is also producing tools to support annotation, linking, search and browsing of content. But a pressing need is to develop standards and methods to describe the knowledge services themselves, and to facilitate the composition of services into larger aggregates and negotiate workflows. Now the research of pervasive computing (which is about devices everywhere; e.g. in everyday artefacts, in our clothes and surroundings, and in the external environment) is more and more in vogue. Pervasive computing provides the manifestation of the grid in the physical world. With the sensor networks as sensors and sensor arrays evolve, we can acquire data with higher temporal or spatial resolution, and this increasing bulk of (often realtime) data demands the computational power of the grid. We need service description, discovery and composition to be applied both to grid and to pervasive computing.

8.3   Semantic web services

Currently technology around UDDI, WSDL, and SOAP provides only limited support in service recognition, service configuration and combination, service comparison and automated negotiation. Thus the ambition for Semantic Web services is to raise the level of description such that services are detailed in a way that indicates their capabilities and task achieving character.

8.3.1   Service description

At present, the Web Services Description Language (WSDL) is already an essential building block in the Web service technologies, and is being developed and standardized in the W3C's Web Services Description Working Group. WSDL, in essence, allows for the specification of the syntax of the input and output messages of a basic service, as well as other details needed for the invocation of the service.

However WSDL lacks semantic descriptions of the meaning of inputs and outputs that makes it impossible to develop software clients that can, without human assistance, dynamically find and successfully invoke a service. Moreover WSDL does not support the specification of workflows composed of basic services. Thus richer semantic descriptions must be added to support greater automation of service selection and invocation, automated translation of message content between heterogeneous inter-operating services, automated or semi-automated approaches to service composition, and more comprehensive approaches to service monitoring and recovery from failure [319].

For the service discovery, we require a semantic language that can be used to encode Web service capabilities for advertisement and for requests. There are two ways to represent functionalities. The first approach provides an extensive ontology of functions where each class in the ontology corresponds to a class of homogeneous functionalities. The second way to represent capabilities is to provide a generic description of function in terms of the state transformation that it produces. Both ways use ontologies to provide the connection between what the Web service does and the general description of the environment in which the Web service operates.

In order to perform automated WS composition, a rich semantic representation of Web service inputs, outputs, preconditions and effects (IOPEs) is needed to resolve constraints between IOPEs. The constraints between these inputs, outputs, preconditions and effects dictate the composition of Web services. For example, we may want to achieve an object (some desired outputs and effects), and match it to the outputs and effects of a Web service (modeled as a process). The result is an instantiation of the process, plus descriptions of new goals to be satisfied based on the inputs and preconditions of that process. The new goals (inputs and preconditions) then naturally match other processes (outputs and effects), so that composition arises naturally. WS preconditions and (conditional) effects are not encoded in any existing industrial standard [319], thus we need a semantic description to encode them in unambiguous computer-interpretable form.

Throughout the presentation of D. Martin and his colleagues [319], most of the work on discovery of Web services using the semantic technologies has been based on OWL-S (Ontology Web Language for Services) which seeks to provide the building blocks for encoding rich semantic service descriptions, in a way that builds naturally upon OWL. OWL-S can help to enable fuller automation and dynamism in many aspects of Web service provision and use, support the construction of powerful tools and methodologies, and promote the use of semantically well-founded reasoning about services.

WSMO handles functional descriptions (WSMO capabilities) to the whole service. It makes an explicit distinction between requester goals and provider capabilities: while capabilities are composed of Preconditions, Postconditions, Assumptions and Effects (plus non-functional properties), goals comprise only Postconditions and Effects. Semantics of Preconditions and Postconditions could be assimilated to IOPE restrictions over Inputs and Outputs. Assumptions and Effects describe the change of service state before and after the service execution, and could be assimilated to IOPR Preconditions and Results [333].

8.3.2 WS-Resources description and shortcomings

Building on concepts and technologies from both the grid and Web services communities, OGSA (Open Grid Services Architecture) defines a uniform exposed service semantics (the grid service); defines standard mechanisms for creating, naming, and discovering transient grid service instances; provides location transparency and multiple protocol bindings for service instances; and supports integration with underlying native platform facilities [321]. The release of WSRF showed that an OGSA infrastructure with stateful WS-

Resources could be built on top of plain WSDL by retaining Web services as stateless entities [333]. The WSRF decouples a Web service (WSDL interface) from its stateful resource (an XML ResourceProperties Document, therefore RPD). Hence, grid researchers can focus on achieving better descriptions, discovering and matching of standard Web services.

As we have said in section 8.3.1, the semantic technologies are adopted in the Web service descriptions to facilitate the service discovery and to construct automatically some complex applications. WSRF deals with application data (resource properties) in the form of XML. WSRF mandates that such application data be described using XML schema and it proposes languages like XPath for querying the data. XML schema together with XPath 2 can express type hierarchies but they lack any semantic reasoning abilities. Research in extending XML schemas with semantics and using some semantic query language (or extending XPath with semantics) can potentially have far-reaching applicability, and the results would be reusable in WSRF as well, because WSRF does not constrain the XML schemas used to describe the data and it allows any dialect of query language. Moreover, by reusing the same results in semantic annotation, understanding and querying of XML data as mentioned above, notification subscription based on semantic matching appears to be a natural combination. Finally, it might be very useful to include resource property availability matching in Web service discovery, i.e. looking for services based on what data they make available as their resource properties, as opposed or in addition to the capabilities of the application-specific service interfaces [326].

Both OWL-S and WSMO have some limitations in describing the WS-Resource [333]. First, OWL-S links IOPE properties solely to WSDL operations, so it is unable to express statements over the service itself. In order to well describe a service, constraints, semantic rules or matchable statements at various levels within an interface are needed. Thus the current descriptions of OWL-S is a limiting factor. Some research suggests allowing WSDL-S pre- and post-elements to appear outside operations; however, these modifications are far from being a general solution. Then, we can emphasize a semantic annotation problem. Currently the final outcome of a semantic annotation process will be a hardwired-at-design-time set of semantic statements as part of the service's profile. But considering the dynamic changes of WSRP value, this is not sufficient to describe well capabilities of WS-Resource. Finally, the point is related with a possible matching and discovery mechanism involving WS-Resources. As we know, a requestor could find a service able to fulfill its requirements by querying a UDDI registry, and then invoking one of the matched endpoint addresses in order to retrieve its WSDL description. Several works have reused this model to advertise semantic service descriptions, enhancing registries with semantic matchmaking algorithms. Again, in order to discover WS-Resource descriptions, the model of UDDI would require a continuous polling of capabilities because of the state changes which can occur at any time.

8.3.3 Semantic WS-Resource description proposals

Currently, some research [333, 316] in semantic-enabled WSRF (WSRF-S) has started and seeks to benefit from adding semantics to the stateful services.

A three-layered Semantic WS-Resource description model is described in [333]. The features of the model are as follows. Firstly, the hereafter Property Layer manages resource properties related to both static characteristics and dynamic changes in the state of the service, conveying the WS-Resource RPD. In turn, WSDL I/O parameters are annotated with semantics as WSDL-S does. Secondly, a Policy Layer is used to transform parameters and properties into explicit capabilities or requirements by attaching policies to each WSDL or RPD element (hence controlling their degree of modality and the type as being functional or non-functional). Finally, the Semantic Topics Layer envelops a concrete set of capabilities/requirements with its policy information, forming a collection of semantic topics (which is the effective profile of the service). However, the current WS-Topics standard covers only syntactic, tree-based matching by means of XPath-like expressions (simple, concrete and full dialects). In order to achieve a distributed sharing of semantic capabilities regarding WS-Resource state changes, WS-Topics should be extended to support semantic topics expressing both statements and Resource Properties.

In the paper [316], the design and development of the semantic grid services for flood forecasting simulations are presented. The authors describe an architecture of the system components of the workflow orchestration and execution environment, and introduce the process of service annotation, discovery and composition in the project K-Wf Grid. The design of the flood-forecasting application services is based on the Web Service Resource Framework (WSRF). A simple ontology is designed to describe resource properties and the association with the instance service and service factory. A WSRFServiceProfile is defined, extending the already existing OWL-S Profile. In order to describe the resource properties, an additional object property is introduced called hasResourceProperty, which points to one or many instances of the class ResourceProperty. Since resource properties are already defined as an XML schema, the ontological description can be created semi-automatically.

8.4 Semantic matching of web services

8.4.1 Matchmaking Systems

InfoSleuth [330] is an agent-based information discovery and retrieval system which adopts "broker agents" to perform the syntactic and semantic matchmaking. The broker agent matches agents that require services with other agents that can provide those services. By saving all the up-to-date information about the operational agents and their services in a repository, the broker enables the querying agent to discover all available agents that provide appropriate services. In InfoSleuth, syntactic brokering is the process of matching requests to agents on the basis of the syntax of the incoming messages; semantic brokering is the process of mapping requests to agents on the basis of the requested agent's capabilities or services, with the agent's capabilities and services being described in a common shared ontology of attributes and constraints. This single domain-specific ontology is a shared vocabulary that all agents can use to specify advertisements and requests to the broker. LDL++, a logical deduction language, is used to describe the service capability information in InfoSleuth. Agents use a set of LDL++ deductive rules to support inferences about whether an expression of requirements matches a set of advertised capabilities.

A multiagent infrastructure, RETSINA (Reusable Task Structure-based Intelligent Network Agents), has been developed by Sycara and his colleagues [335]. They distinguished three general agent categories in Cyberspace: service provider, service requester, and middle agent. They define Matchmaking as the process of finding an appropriate provider for a requester through a middle agent. To describe these agents' capabilities in the matchmaking process, they have defined and implemented an ACDL (Agent Capability Description Language), called Larks (Language for Advertisement and Request for Knowledge Sharing). Larks offers the option to use application domain knowledge in any advertisement or request by using a local ontology, written in a specific concept language ITL, to describing the meaning in a Larks specification. The matching process uses five different filters: context matching, profile comparison, similarity matching, signature matching and constraint matching. Different degrees of partial matching can result from utilizing different combinations of these filters. The selection of filters to apply is under the control of the user or the requester agent.

8.4.2   Matching engine

The matching process is achieved in the component named Matching engine. The core problem is the matching algorithms which will be discussed in 8.4.3. Another problem is how a *sufficiently similar* service can be found to match a request. The problem is the definition which specifies what *sufficiently similar* means. To accommodate a softer definition of *sufficiently similar* we need to allow matching engines to perform flexible matches [332]. Matches recognize the degree of similarity between advertisements and requests. Service requesters should also be allowed to decide the degree of flexibility that they grant to the system.

The matching engine must satisfy the following characteristics [332]:

- The matching engine should support flexible semantic matching between advertisements and requests on the basis of the ontologies available to the services and the matching engine

- The requesting service should have some control on the amount of matching flexibility it allows to the system

- The matching engine should encourage advertisers and requesters to be honest with their descriptions

- The matching process should be efficient: long delay for the requester is not tolerated

8.4.3 Semantic matching algorithms

Normally, when all the outputs of the request are matched by the outputs of the advertisement, and all the inputs of the advertisement are matched by the inputs of the request, we define that an advertisement matches a request. This criteria guarantees that the matched service satisfies the need of the requester.

Some matching algorithms are presented in [332, 328]. The main control loop of the matching algorithm is shown in algorithm 8.4.1. When a request arrives, it is matched with all the advertisements recorded by the registry. Whenever a match between the request and any of the advertisements is found, the match is recorded and sorted at the end of this loop.

Algorithm 8.4.1 The main control loop

recordMatch = empty list
for Each advertisement in register **do**
  if advertisement matches request **then**
    recordMatch.append(request, advertisement)
  end if
end for
sort(recordMatch)

A match between an advertisement and a request consists of the match of all the outputs of the request against the outputs of the advertisement and the match of all the inputs of of the advertisement against the inputs of the request. The algorithm for output matching is presented in algorithm 8.4.2. A match is recognized if and only if for each output of the request, there is a matching output in the advertisement. The degree of success will be returned when the match loop is finished. If one of the request's outputs is not matched by any of the advertisement's output, the match fails. The matching between inputs is similar, but with the order of the request and the advertisement reversed.

Algorithm 8.4.2 The Algorithm for output matching

---

globalMatchDegree = exact
for Each outR ∈ outputRequest **do**
    matchDegree = maxMatchDegree(outR, outA ∈ outputAdvertisement)
    if matchDegree = fail **then**
        return fail
    end if
    if matchDegree <globalMatchDegree **then**
        globalMatchDegree = matchDegree
    end if
end for
return globalMatchDegree

There are four degrees of matching according to [332]. The description of these matching degrees is presented below.

exact If advertisement A and request R are equivalent concepts, we call the match exact; formally, $A \equiv R$.

plug in If request R is a sub-concept of advertisement A, we call the match plugIn; formally, $R \sqsubseteq A$.

subsume If request R is a super-concept of advertisement A, we call the match subsume; formally, $A \sqsubseteq R$.

fail Failure occurs when no subsumption relation between advertisement and request is identified.

8.5 Semantic workflow

In order to satisfy client requirements, Web services will be composed as part of workflows to build complex applications. As we have described in Chapter 7, several workflow specification languages have been developed to achieve the workflow composition. But the workflow composition techniques require the user to deal with low-level details and the process of composition is normally manual or semi-manual. Automated composition techniques can be used to automate the entire composition process by using AI planning or similar technology.

8.5.1 Model for composing workflows

Mikko Laukkanen and Heikki Helin [327] describe a model for composing web service workflows by utilizing semantic Web service ontologies. Figure

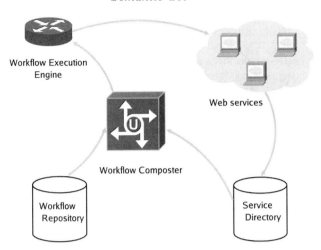

FIGURE 8.2: Architecture of model for dynamic workflow composition.

8.2 depicts the general entities of the web service composition model and the information flow between them. The capacity of Web services is described by a semantic description language, such as DAML-S. When a new web service instance is created, it is advertised by registering the WSDL and DAML-S description to the directory, which can be for instance UDDI. The workflows are defined in a workflow specification language and are stored in the workflow repository which can be used by the workflow composer agent. Each workflow is composed of one or more Web services, which can be situated anywhere on the web. Then the workflow composer agent can query the DAML-S and WSDL descriptions of the web services from the service directory, and the semantic matching is applied. The executable workflow is composed by the workflow composer agent and is fed into the workflow execution engine. Finally, the execution engine executes the workflow using the web service instances.

The paper [329] presents an architecture to facilitate automated discovery, selection, and composition of semantically described heterogeneous services using Semantic Web technologies. Three main features distinguish the framework from other work in this area. First, a dynamic, adaptive, and highly fault-tolerant service discovery and composition algorithm is proposed. Second, a distinction between abstract and concrete workflows is made. This facilitates the workflow share in the system. Finally, the framework allows the user to specify and refine a high-level objective.

The framework architecture is shown in Figure 8.3. The WFMS is the co-ordinator of the entire process and manages the flow of messages between the components. The AWFC service generates an abstract workflow according to the incoming request. The AWFC will typically query the AWFR to ascertain

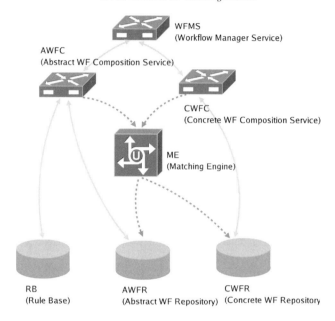

FIGURE 8.3: Architecture of the framework.

if the same request has been processed previously. If so, the abstract work-flow will be returned to the Manager Service. If not, a request will be made to the ME to retrieve a process template from the RB which can satisfy the request. If a process template is not available, an attempt will be made to retrieve a combination of tasks that provide the same functionality based on the inputs, outputs, preconditions and effects. Then an abstract workflow is generated and returned to the Manager Service. The CWFC service achieves the mapping of an abstract workflow with available instances of actually de-ployed services on the network at that time. Here the problem is that a service implementation may not be available. If the matching process is successful, an executable graph is generated and returned to the Manager Service. If not, the AWFC is invoked through the WFMS, and asked to provide an alternative abstract workflow. This provides a high degree of fault tolerance.

8.5.2 Abstract semantic Web service and semantic template

For the purpose of service composition, we normally focus on the abstract representation of Web services, i.e. operations and messages, but do not con-sider the binding detail. A definition of abstract semantic Web service is presented in [337]. An abstract semantic Web service, SWS, can be repre-sented as a vector:

$$SWS = (sop1, sop2, ..., sopn)$$

Each *sop* is a semantic operation defined as a 6-tuple:

$$sop = < op, in, out, pre, eff, fault >$$

Each tuple is a semantic description of a Web service's property. *op* represents the operation, *in* is the input message, *out* is the output message, *pre* represents the precondition, *eff* is the effect and *fault* is the exceptions of the operation represented using classes in an ontology.

For a service provider, we use an abstract semantic Web service definition to represent the operations and messages of a service. In the same way, a semantic template is used to model the requirement of the service requester. It is the way a service requester models the data, functional and non-functional specifications of a task.

A semantic template (ST) can also be represented as a vector:

$$ST = (sopt1, sopt2, ..., soptn)$$

Each *sopt* is a semantic operation template, which is defined as a 6-tuple:

$$sopt = < op, in, out, ssf0, gl, fault >$$

- *op* is the semantic description of the operation template.

- *in* is the semantic description of the initial message.

- *out* is the semantic description of the output message.

- *ssf0* is the semantic description of the initial status flags.

- *gl* is the semantic description of the goal.

8.5.3 Automatic Web service composition

Two types of automatic Web service composition are described in [336], complete automation and full automation.

Complete automation can be achieved when the execution requirements of the workflow components are specified, and the system can query the execution environment to find what resources are available for execution. In this case, optimizing the completion time of any given workflow, considering resource assignment trade-offs across many workflows, and designing appropriate failure handling and recovery mechanisms could be important challenges. The complete automation can also be applied to complete underspecified workflow templates. Workflow templates are underspecified when they include abstract computation descriptions to be specialized during workflow instance creation. The binding of abstract description with an instance of service may depend on the nature of the workflow's processes. Workflow templates can also be underspecified in that they may be missing workflow components that perform no critical and standard processing in the workflow. Automatically adding these steps is possible when the component library includes appropriate components for doing the kinds of processing required. The kinds of processing needed may not be known until the workflow instance is created and therefore would not typically be included in a workflow template. Once initial status is specified, new processing steps which haven't been specified in the workflow template can be added during workflow instance creation.

Full automation of the workflow composition process may be desirable for some kinds of workflows and application domains. Given a description of the desired data products, a valid workflow can be created from individual application components that provide full-fledged automatic programming capabilities. In order to achieve automatic workflow generation, application components must be clearly encapsulated and described with detailed specifications of the component's outputs based on the properties of their input data. These specifications must include criteria for component selection and data selection when several alternatives are appropriate. As an alternative to creating new workflows from scratch, fully automatic workflow generation can also be achieved by reusing workflow templates. This approach requires a library of workflow templates that satisfies the common requirement in the domain and can be queried according to the key characters of workflow template, such as the initial message or the goal.

8.6   Concluding remarks

The convergence of semantic technologies and grid computing provides lots of advantages. The integration of semantic technologies into Web service raises the level of description such as their capabilities and task achieving character. Thus this integration provides the support in service recognition, service configuration, service comparison and automated composition. The key technologies for semantic service description have been heavily studied. But a pressing need is to develop standards and methods to describe the knowledge services themselves, and to facilitate the composition of services into larger aggregates and negotiate workflows.

References

[316] M. Babik and M. Maliska. Semantic grid services in k-wf grid. In *Proceeding of the 2nd International Conference on Semantics, Knowledge and Grid*. IEEE Press, 2006.

[317] M. C. Brown. What is the semantic grid, September 2005. Available online at: `http://www-128.ibm.com/developerworks/grid/library/gr-semgrid/` (Accessed 30th September, 2007).

[318] M. Cannataro and D. Talia. Semantics and knowledge grids: Building the next-generation grid. *IEEE Intelligent Systems*, 19(1):56–63, 2004.

[319] D. Martin et al. Bringing Semantics to Web Services: The OWL-S Approach. In *First International Workshop on Semantic Web Services and Web Process Composition (SWSWPC 2004)*, 2004.

[320] J. Euzenat. Towards formal knowledge intelligibility at the semiotic level. In *ECAI 2000 Workshop Applied Semiotics: Control Problems*, pages 59–61, 2000.

[321] I. Foster, C. Kesselman, J. Nick, and S. Tuecke. Grid services for distributed system integration. *Computer IEEE*, 35:37–46, 2002.

[322] Y. Gil, E. Deelman, J. Blythe, C. Kesselman, and H. Tangmunarunkit. Artificial intelligence and grids: Workflow planning and beyond. *IEEE Intelligent Systems*, 19(1):26–33, 2004.

[323] T. R. Gruber. A translation approach to portable ontology specifications. *Knowledge Acquisition*, 5(2):199–220, 1993.

[324] I. Horrocks. DAML+OIL: a description logic for the semantic web. *IEEE Data Engineering Bulletin*, 25(1):4–9, 2002.

[325] J. Hunter, J. Drennan, and S. Little. Realizing the hydrogen economy through semantic web technologies. *IEEE Intelligent Systems*, 19(1):40–47, 2004.

[326] J. Kopeck. Semantic web services resource framework (wsrf-s) report. Technical report, DERI, 2005. WSMO Working Draft 8 August 2005.

[327] M. Laukkanen and H. Helin. Composing workflows of semantic web services. In *Proceedings of the Workshop on Web-Services and Agent-based Engineering*, 2003.

[328] L. Li and I. Horrocks. A software framework for matchmaking based on semantic web technology. In *Proceedings of the Twelfth International World Wide Web Conference (WWW 2003)*, 2003.

[329] S. Majithia, D. W. Walker, and W. A. Gray. Automated web service composition using semantic web technologies. In *First International Conference on Autonomic Computing (ICAC'04)*, 2004.

[330] M. H. Nodine, J. Fowler, T. Ksiezyk, B. Perry, M. Taylor, and A. Unruh. Active information gathering in InfoSleuth. *International Journal of Cooperative Information Systems*, 9(1–2):3–28, 2000.

[331] S. B. Palmer. The semantic web: An introduction, September 2001. Available online at: http://infomesh.net/2001/swintro/ (Accessed 30th September, 2007).

[332] M. Paolucci, T. Kawamura, T. R. Payne, and K. P. Sycara. Semantic matching of web services capabilities. In *ISWC '02: Proceedings of the First International Semantic Web Conference on The Semantic Web*, pages 333–347, London, UK, 2002. Springer-Verlag.

[333] R. Rodriguez, C. Costilla, and A. Calleja. A Topic-based approach to express dynamic capabilities of Semantic WS-Resources. In *AICT-ICIW '06: Proceedings of the Advanced Int'l Conference on Telecommunications and Int'l Conference on Internet and Web Applications and Services*, page 181, Washington, DC, USA, 2006. IEEE Computer Society.

[334] D. D. Roure, N. R. Jennings, and N. R. Shadbolt. The semantic grid: Past, present, and future. In *Proceedings of the IEEE*, volume 93, pages 669–681, March 2005.

[335] K. Sycara, J. Lu, M. Klusch, and S. Widoff. Dynamic service matchmaking among agents in open information environments. *ACM SIGMOD Record (Special Issue on Semantic Interoperability in Global Information Systems)*, 28(1):47–53, 1999.

[336] I. J. Taylor, E. Deelman, D. B. Gannon, and M. Shields, editors. *Workflows in e-Science*, chapter Workflow Composition: Semantic Representations for Flexible Automation. Springer Verlag, 2006.

[337] Z. Wu, A. Ranabahu, K. Gomadam, A. P. Sheth, and J. A. Miller. Automatic composition of semantic web services using process and data mediation. Technical report, LSDIS Lab, University of Georgia, February 2007.

Chapter 9

Integration of scientific applications

9.1 Introduction

A perennial problem with grid applications is making them flexible enough to be used across a range of potential platforms and environments. However, minor differences between platforms can cause significant problems. For example, changes between Windows versions, even Windows NT and Windows 2000, can cause problems for such rigidly designed and optimized applications as are usually employed in a grid environment. An obvious solution is to remove the highly platform-specific elements and move to a more generalized environment. By making the grid application run on a wider range of platforms, this generalized environment enables easy expansion of the scope and power of the grid simply by adding more machines [340].

The SOA (service-oriented architecture) is a component-based model for building applications that divides applications into a number of discrete services that, individually, perform a specific function, but when put together make up the components of a larger application. By making Web services easier to find and identify, SOA makes it easier to deploy and distribute an SOA-based application. Because Web services are based on open standards and are, by definition, architecture- and platform-neutral, SOA-based applications can be deployed across a wide range of platforms.

In short, SOA is a method for exposing services and allowing computers to talk to each other and share power and functionality. Grids have slowly been moving toward a Web services architecture, first with the move by Globus to the Open Grid Standards Infrastructure (OGSI) and, further, with the Globus Toolkit 4.0 (GT4) release. SOA and grid technology are moving toward the Web Standards Inter-operability technology, based on solutions such as the Web Services Resource Framework (WSRF) and others.

Many scientific communities are feeling a growing need to convert their legacy applications into Web services. Unfortunately, most of the applications developed and used by scientific communities are command-line applications written in FORTRAN, C and a host of scripting languages. They are fast, efficient and easy to use. However, they are usually platform dependent and are difficult to integrate with applications from other communities. There is no standard way of registering these applications so that they can be discovered by interested clients and end-users. Also, there is no standard way to describe their input parameters and output results and to monitor their progress as they run for extended periods of time on the grid. By converting these command-line applications into grid services, most of the aforesaid limitations can be overcome [357, 371].

The primary goal of this chapter is to implement a framework for dynamic deployment of scientific applications where the end-users can:

- Apply any legacy code as a WSRF-compliant service when they create grid applications.

- Deploy dynamically any scientific application into the grid environment.

- Utilize a uniform interface to interact with any deployed application.

In the framework, the scientific applications are described as job description files in XML format [365]. We utilize the WSRF resource [366] to contact a local job manager through Globus [349] to submit the legacy computational job. The factory service manages all these job descriptions and creates the resource according to the client request. The instance service supplies a uniform interface for all applications. This interface is used to submit and to monitor the applications. Our framework has four primary components:

- *Factory service* that manages all the application descriptions and returns a list of applications to the client interested. It also has a mechanism to monitor the creation, deletion, and modification of the application description. Thus we can dynamically make some applications available or unavailable on the grid. According to the selected application by the client, the *Factory service* creates a resource and returns an endpoint reference composed of the *Application service* and the recently created resource to the client.

- *Application service* that provides a uniform interface for the client to invoke the applications in the computing resource and to monitor the status of application executions.

- *AdminTool* which can interact with the *Factory service* in a secure way. The *AdminTool* has a graphic interface and can be used to add, delete and modify the application descriptions by the local administrator.

- *Application Scheduler* is a meta-scheduler in our framework. The *Application Scheduler* manages and monitors all available computing resources in a VO [345]. According to the request of the client, it interacts with the *Factory service* in each computing resource to get the applications list, collects the dynamic and static information of computing resources to make a scheduling decision, invokes the *Factory service* in the computing resource to create a WSRF resource for the user, submits applications, and monitors the execution status.

In each available computing resource, only one *Factory service* and one *Application service* run persistently. No other instance service is created. So the creation and the management of an application instance are standard and simple.

The rest of this chapter is as follows. In Section 2 we discuss some frameworks which can be used to achieve the application integration into the grid. Then in Section 3 the model architecture and the implementation are described. In Section 4 we provide more details about the application description and the service creation. The security issue is discussed in section 5. At last, we conclude with a brief discussion of future research.

9.2 Framework

There are several research efforts aiming at automating the transformation of legacy code into a grid service. Most of these solutions are based on the general framework to transform legacy applications into Web services outlined in [359], and use the standard method of Web service to discover these services, submit jobs, and monitor the execution status. The other solution is Java wrapping which can generate stubs automatically for legacy applications. One example can be found in [355], where the authors describe a semi-automatic conversion of legacy C code into Java using JNI (Java Native Interface) [356]. A solution based on WSRF is also described in this section.

9.2.1 Java wrapping

The paper [355] describes a process for the semi-automatic conversion of numerical and scientific routines written in the C programming language into Triana-based computational services that can be used within a distributed service-oriented architecture of grid computing. This process involves two separate but related tools, JACAW and MEDLI. JACAW is a wrapper tool based on the Java Native Interface (JNI) that can automatically generate the Java interface and related files for any C routine, or library of C routines. The MEDLI tool can then be used to assist the user in describing the mapping between the Triana and C data types involved in calling a particular routine.

9.2.2 Grid service wrapping

Compared to Java wrapping GEMLCA [356] is based on a different principle. It offers a front-end grid service layer that communicates with the client in order to pass input and output parameters, and contacts a local job manager through Globus MMJFS (Master Managed Job Factory Service) to submit the legacy computational job. There is no need for the source code and not even for the C header files to deploy a legacy application as a grid service. The legacy code can be written in any programming language and can be a sequential or a parallel PVM or MPI code that uses a job manager like Condor. The current implementation of GEMLCA is based on GT (Globus Toolkit) but the architecture itself is more generic and can be easily adapted to other service-oriented approaches like WSRF or a pure Web services-based solution. To wrap the legacy applications, the user only has to describe the legacy parameters in a pre-defined XML format that in the current GEMLCA version has to be done manually. However, the next release of the architecture will automate this process.

The paper [357] presents a framework that allows scientists to wrap their applications as services, deploy them on the grid, securely interact with these services, compose scientific workflows using these services and monitor the status of their workflows on the Grid. The framework has four primary components:

- A grid portal, which is a Web server and a gateway for users to access services, compose workflows and manage data.

- A generic Factory Service that is invoked from the Portal by application providers to wrap applications as services and create new instances of these services on the grid.

- A workflow composer tool that allows users to compose complex and interesting workflows from application services.

- A Notification Service that allows application services to send messages that are logged by the Portal and monitored by the workflow instance.

The services created by this framework generate their own graphical user interface, which allows end-users to interact with them using thin and generic Web service clients. This framework takes care of authentication and authorization during all client-service interactions in a manner transparent to application developers and end-users.

An implementation of an Application Factory Service is described in [346]. The Factory Service is designed to create instances of distributed applications that are composed of well-tested and deployed components each executing in a well-understood and predictable hosting environment. The basic technology used to build such a factory service is based on XCAT, which is a grid-level implementation of the Common Component Architecture developed for the U.S. Department of Energy. XCAT can be thought of as a tool to build distributed-application-oriented web-services. The application factory service (AFS) is a stateless component that can be used to launch many types of applications. This Factory Service accepts requests from a client that consists of static and dynamic application information. The AFS authenticates the request and verifies that the user is authorized to make the request. Once authentication and authorization are complete, the AFS launches an application coordinator. Then AFS passes the static and dynamic application information to the application coordinator to create the application component. Once the application component instances have been created, connected and initialized, the application coordinator builds a WSDL document of the ensemble application and returns that to the AFS, which returns it to the client. The factory service maintains no state information about any of the application instances it launches.

9.2.3 WSRF resources

In the grid services wrapping, the interface by which we can interact with the deployed applications is not uniform. Because the Factory needs a description of the service to create an instance of application, the different description providers could define various service port-types in the descriptions. Therefore the interface of application instances varies according to different service port-types. The other problem is the quantity of service instances. The application is created and deployed as a service instance. In this case, if we deploy a large quantity of needed applications in a computing resource, there will be too many service instances to be created. The management of these instances is truly a delicate job.

The framework WSRF provides a solution to solve these problems. WSRF separates service status from grid service and maintains the status in WSRF resources. Thus, we can use WSRF resources to wrap legacy applications and make grid services the interface which is used to submit the computational applications and to monitor the execution status. With this solution, we normally deploy one factory service to create WSRF resources and one grid service to interact with users in each computing resource. The management of services is straightforward and the interface to be used by users could be unique and standard.

9.3 Implementation

The framework presented in this chapter adopts the model of SOA and each component in the framework is wrapped into Web service and is developed on top of GT 4 (Globus Toolkit). So before discussing the implementation of the framework, we will introduce some basic concepts of the Globus Toolkit.

9.3.1 Globus Toolkit and GRAM

The Globus Toolkit (GT) was developed in the late 1990s to support the development of service-oriented distributed computing applications and infrastructures. The Web services-based GT4 is the latest release of GT, which provides significant improvements over previous releases in terms of robustness, performance, usability, documentation, standards compliance, and functionality [342].

The GT4 Grid Resource Allocation and Management (GRAM) service addresses the issues of running a task on a computer, providing a Web services interface for initiating, monitoring, and managing the execution of arbitrary computations on remote computers [342]. The Globus Toolkit provides both a suite of web services and a "pre-web services" Unix server suite to submit, monitor, and cancel jobs on grid computing resources. Both systems are called "GRAM", while "WS GRAM" refers only to the web service implementation. In GT4, jobs are computational tasks that may perform input/output operations and the execution of jobs affects the state of the computational resource and its associated file systems. In practice, such jobs may require coordinated staging of data into the resource prior to job execution and out of the resource following execution.

Grid computing resources are typically operated under the control of a scheduler which implements allocation and prioritization policies while optimizing the execution of all submitted jobs for efficiency and performance. GRAM is not a resource scheduler, but rather a protocol engine for communicating with a range of different local resource schedulers using a standard message format [348].

Other services and systems can be needed when we submit a task to GRAM. The Globus Toolkit's Monitoring and Discovery System (MDS) defines and implements mechanisms for service and resource discovery and monitoring in distributed environments [364]. The Grid Security Infrastructure (GSI) [350] is the portion of the Globus Toolkit that provides the fundamental security services needed to support grids. It also provides a number of components for data management such as the Globus GridFTP tools and the Globus Reliable File Transfer (RFT) service [347].

9.3.2 Architecture and interface

Figure 9.1 illustrates the architecture of the model. A *Grid Resource* is a computing resource on which GT4 has been installed and on which the *Factory service* and *Application service* have been run persistently. We have one *Application Scheduler* running as a meta-scheduler of the VO. The meta-scheduler is the grid portal for clients and it manages all the *Grid Resource*s in the VO. It interacts with the *Factory service* and *Application service* to create resources, submit computational jobs and monitor the jobs status for clients.

MDS can be configured in a hierarchical fashion with upper levels of the hierarchy aggregating information from the lower-level MDS (Index Services). The upper levels are identified as upstream resources in the hierarchy, and the lower levels are identified as downstream resources [362]. Thus from the local MDS, the *Application Scheduler* can gather the dynamic and static

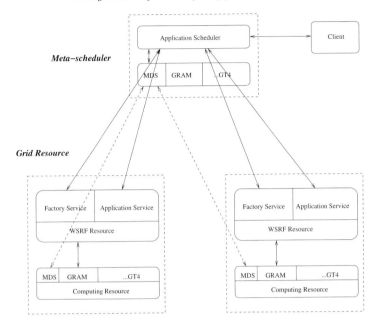

FIGURE 9.1: The architecture of the proposed model.

information from each *Grid Resource* in the VO.

The architecture of the *Grid Resource* is shown in Figure 9.2. An application storehouse stores the application descriptions which support the Job Description Schema [351]. An *AdminTool* interacts with the *Factory service* to add, delete and modify application descriptions. According to the request of the *Application Scheduler*, the *Factory service* can create a resource and submit a computational job for the user. The resources use GRAM to actually submit a job to the computing resource and subscribe to the notification of job status [352] to monitor the job execution [366]. The information of application execution is stored inside the resource and, more specifically, in resource properties.

But how can the user set the arguments and stage files of the application? In the Job Description Schema, we have three elements: *Argument, FileStageIn* and *FileStageOut* [351]. After a *Grid Resource* has been selected by the *Application Scheduler*, the user specifies all the input parameter values (including *Argument, FileStageIn* and *FileStageOut*) and sends a submission request to the *Application Scheduler*. Then the *Application Scheduler* sets these elements in the Job Description and invokes the operation *createResource* of *Factory service* with the Job Description as the parameter. The Factory service uses the Job Description to initialize the resource.

Table 9.1: The PortType of Services

The PortType of Application Scheduler		
	PortType	Description
1	openSession	open a session for user
2	closeSession	close the user session
3	findApplication	search the application in the *Grid Resource*. If there is more than one available *Grid Resource*, we use MDS information to select the best resource for user
4	scheduler	submit the application to *Grid Factory*
5	getJobStatus	return the job execution status

The PortType of Factory Service		
	PortType	Description
1	getApplicationList	return a list to client
2	createResource	create resource for client
3	addApplication	add Job Description
4	modifyApplication	modify Job Description
5	deleteApplication	delete Job Description

The PortType of Application Service		
	PortType	Description
1	submit	invoke operation *submit* of resource to submit the job to GRAM
2	stop	stop the job execution
3	getJobStatus	get job status from resource

Based on the GT4 and WSRF, we realize our *Grid Scheduler, Factory service* and *Grid service*. The PortType [369] of each service is illustrated in Table 9.1.

9.3.3 Job scheduling and submission

In the *Application Scheduler*, we implement a simple scheduling algorithm. When the *Application Scheduler* finds that there are more than one available *Grid Resource* for the user, it compares the number of available CPUs of each *Grid Resource*. The *Application Scheduler* selects the resource which has the most available CPUs. If the number of available CPUs is similar, the *Application Scheduler* calculates the value of $Waitingjobs/TotalCPUs$ for each *Grid Resource*. $Waitingjobs$ is the number of jobs waiting in the local job queue, and $TotalCPUs$ is the number of CPUs on each *Grid Resource*. The resource which has the smallest value is selected. A more complex scheduling algorithm will be considered in the future.

In the framework, it is the resource that contacts a local job manager through Globus to submit the computational job. Figure 9.3 shows how the resource is created.

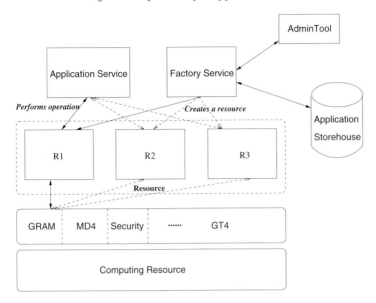

FIGURE 9.2: The Architecture of Grid Resource.

1. *Application Scheduler* needs to know only the URI (Uniform Resource Identifiers) of the *Factory service*. With this URI, it can invoke the *getApplicationList* operation. This will return a *String* containing the list of all available applications.

2. *Application Scheduler* selects the application needed for the user. After the user has provided the arguments and stage files of the application, the *Application Scheduler* fills the fields in the Job Description and invokes the *createResource* operation with the Job Description as the parameter. This will return an endpoint reference containing the URI of the Grid Service, along with the key of the recently created resource.

3. The Factory service uses a class *ResourceContext* to get the *GridResourceHome*.

4. The resource home can be used to create the new resource. The creation method returns an object of type *ResourceKey*. This is the resource identifier which is needed to create the endpoint reference that is returned to the *Application Scheduler*.

5. The resource home takes care of actually creating a new resource instance and initializes the resource with the Job Description.

6. The resource home adds the new resource instance to its internal list of resources. This list allows us to access any resource with the resource identifier.

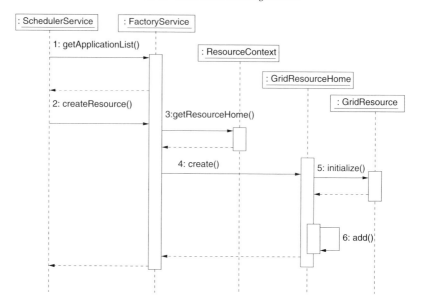

FIGURE 9.3: The sequence diagram for resource creation.

Once the *createResource* call has finished, the *Application Scheduler* will have the WS-Resource's endpoint reference for the user. In all future calls, this endpoint reference will be passed along transparently in all the invocations. So, let's take a close look at what happens when the *Application Scheduler* invokes the *submit* operation, as shown in Figure 9.4.

1. The *Application Scheduler* invokes the *submit* operation in the *Application Service*.

2. The *Application Service* uses *ResourceContext* to retrieve a resource. It will be in charge of reading the EPR and finding the resource.

3. The *Application service* invokes *submitInternal* operation in the GridResource.

4. The GridResource uses GRAMClient [365] that is a Custom GRAM Client for GT4 to actually submit the application.

5. After the submission of the application, the GridResource subscribes to the Notification of job status, and then it can receive a notification when the job status changes and it keeps the job status in the form of a resource property.

6. If the *Application Scheduler* wants to consult the status of a running job, it invokes the *getJobStatus* operation of the *Application service*.

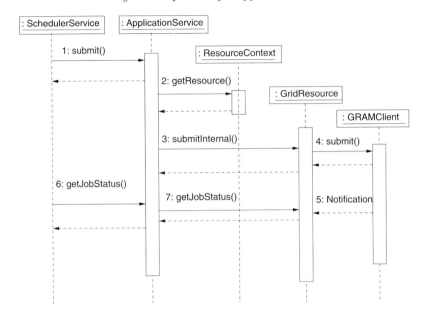

FIGURE 9.4: The sequence diagram for the execution of an application.

7. The *Application service* invokes the *getJobStatus* operation in the GridResource to retrieve the job status.

Figure 9.5 illustrates the sequence of a user job submission.

1. The user invokes the *openSession* operation of the *Application Scheduler* to get a client number.

2. The user invokes the *findApplication* operation with client number and the requested application as parameters.

3. The *Application Scheduler* searches in all the application lists. If it finds the requested application, a Boolean "true" is returned to the user.

4. The user gets "true", so it can invoke the *scheduler* operation in order to submit the application.

5. The *Application Scheduler* invokes *createResource* of the *Factory Service* to create a resource for the user.

6. After having created the resource, the *Application Scheduler* submits the job to *Application Service*.

7. The user uses *getJobStatus* to query the job status.

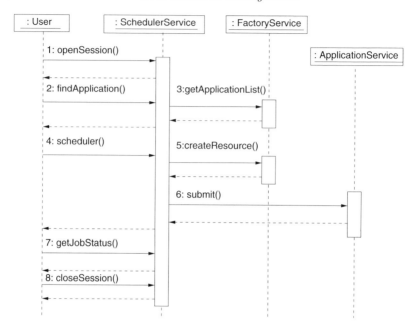

FIGURE 9.5: The sequence diagram for an user job submission.

8. If the execution of the application is finished, the user invokes *closeSession* to destroy the session.

In the *Grid Scheduler* and *Factory Service*, a mechanism is integrated to detect the modification of application descriptions. When the local administrator uses the *AdminTool* to add, delete and modify the application descriptions, the operations (*addApplication*, *modifyApplication* and *deleteApplication*) of the *Factory Service* are invoked. The *Factory Service* then updates the application list and modifies the job description files in the application storehouse. It also sets a signal to notify the *Grid Scheduler* of modification of the application list. The *Grid Scheduler* monitors the signal status. When it detects the change of signal status, it updates its application lists within a reasonable delay.

9.3.4 Code deployment

The AdminTool is used to create the application description by the local administrator of each Grid Resource. Figure 9.6 shows the graphic interface of AdminTool. There are already two applications which are deployed in this Grid Resource and the administrator can use buttons in the toolbar to add, modify and delete applications. If the administrator wants to deploy an application, he clicks on the button in the most left of the toolbar. Then an

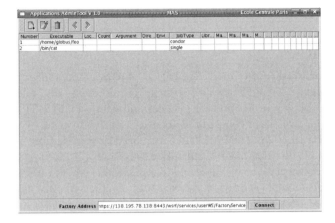

FIGURE 9.6: The graphic interface of AdminTool.

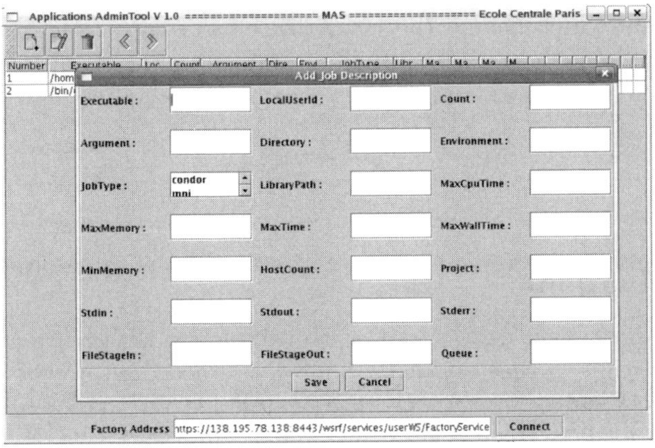

FIGURE 9.7: The dialog to add a job description.

Add Job Description dialog (Figure 9.7) appears and the administrator can fill some or all of these fields in the dialog to create the application description. Then an application description file of this application is transferred to the Factory Service and the Factory Service saves this file to the local application storehouse.

9.4 Security

In any networked environment, security is of paramount concern. GSI is the GT4 component that addresses all security requirements and allows privacy, integrity, and replay protection for grid communication [367]. The framework deals with the two basic concepts of security: authentication (verifying that users are who they say they are) and authorization (assigning privileges to users once their identity has been firmly established).

To enforce security on the client-side, applications which interact with *Application Scheduler*, *Application Service* and *Factory Service* must be configured to use host authorization and to enforce both privacy and integrity of authentication. On the server-side, authentication and authorization are specified by creating a security descriptor file before services (e.g., *Application Scheduler*) are compiled into GAR files [367]. The Gridmap authorization is adopted instead of host authorization on the server-side [372].

User Applications which have the authorization of *Application Scheduler* can interact with the Application Scheduler service. If a User Application submits a job via the *Application Scheduler*, the *Application Scheduler* uses a user account which has all the authorizations of each *Resource Service* to actually submit the job. This mutual authentication mechanism enforces the security of the framework and reduces the complexity of configuration.

9.5 Evaluation

The most important aspect for the job submission is the turn-around time. Turn-around time is the time from a job being accepted by the *Application Scheduler* or *Factory Service* until the completion (i.e., the job has reached the done state). The turn-around time is measured in 2 cases:

- An application is added dynamically in a *Grid Resource*

- The *Factory Service* and *Application Service* are used directly to submit a job without the *Application Scheduler*

9.5.1 Dynamic deployment experiments

As discussed in Section 3, the application can be added dynamically in the system. Thus at first the performance of dynamic deployment is measured. The experimental setup is as follows. The *Factory Service* and *Application Service* are deployed and tested at two Condor clusters: a cluster named $C1$ with three servers, another cluster named $C2$ with two servers. Each server has 2 Pentium 4 3.20GHz with 1 GB RAM. The *Application Scheduler* is installed in a PC powered by Pentium 4 3.00GHz with 512 MB RAM. All the machines are connected by 100 Mb Ethernet. GT 4 is installed in the central manager of Condor pool, and Scheduler Adapters are configured to support the job submission into the Condor pool.

From a laptop, the user submits 30 jobs to the *Application Scheduler* and the interval of submission is 30 seconds. In the user's opinion, a job is a sequence of *openSession*, *findApplication*, *scheduler*, *getJobStatus* and *closeSession*. At the beginning, the application which the user needs is deployed on $C1$. The application is a simple C program. It waits 5 minutes and then returns. In order to execute the application in the standard universe, *condor_compile* must be used to relink the application with the Condor libraries [341]. After the user has submitted 8 jobs, the local administrator of $C2$ runs *AdminTool* to add the application in $C2$. For comparison, the user submits 30 jobs once again. The difference from the first time is that there is not a dynamic deployment.

Figure 9.8 shows that the turn-around time of followed jobs dropped down when the application is added in $C2$ (after the eighth job). Because the *Application Scheduler* detects the modification of applications list in $C2$ and it can submit the user job to $C2$. Thus the ninth job does not wait to be submitted to $C1$; instead it is submitted to $C2$ and is executed immediately. Since the system MDS takes time to gather resource information, the *Application Scheduler* uses the information a little delayed to schedule the jobs. When the fifteenth job is submitted, the *Application Scheduler* submits continually the job to $C2$, because the *Application Scheduler* thinks that there are still some free CPUs in $C2$. This is the reason why the turn-around time of the fifteenth job is a little longer. After the submission of the fifteenth job, the turn-around time of the following jobs in the case of dynamic deployment is much less than in the case of the absence of dynamic deployment because of the distribution of jobs on two clusters.

9.5.2 Grid resource experiments

The *Grid Resource* is the Computing Resource where the *Factory Service* and *Application Service*, called *User Service*, are deployed. Globus provides a standard interface for communicating with Condor using a standard message format. Similarly the *User Service* is deployed on Globus to provide a uniform interface for the job submission. Jobs are submitted separately to the *User Service* and Globus in order to evaluate the performance of the *User Service*.

In these experiments, the application used is a simple MPI program (in C). It calculates parallel the value of Pi using numerical integration in two machines. In order to execute the application in the MPI universe in Condor, the program to be submitted for execution under Condor will be compiled using *mpicc* [341].

All the experiments are done on $C2$. In order to execute parallel applica-

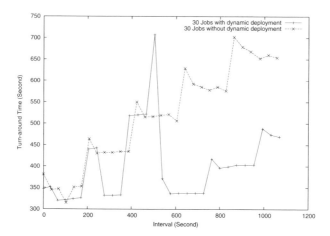

FIGURE 9.8: The performance of submission with meta-scheduler.

tions, MPICH (version 1.2.4) [338] is installed on each server of $C2$. From a laptop, a program submits separately 10, 30 and 50 jobs to local *User Service* with an interval of submission of 5 seconds. Then the Globus command *"globusrun − ws"* is used to submit jobs. The command submits also 10, 30 and 50 jobs with the same interval.

Figure 9.9 shows the result. It is shown that the average turn-around time of User Service is a little longer than the time of *"globusrun − ws"*, except in the case of 30 jobs. The performances of the two infrastructures are very close.

9.6   Concluding remarks

The framework for dynamic deployment of scientific applications into a grid environment has been described. The framework addresses dynamic applications deployment. The local administrator can dynamically make some applications available or unavailable on the *Grid Resource* without stopping the execution of the Globus Toolkit Java Web Services container. An *Application Scheduler* has been integrated in the framework, which can realize simple job scheduling and select the best *Grid Resource* to submit jobs for the users. The performance of the framework has been evaluated by some experiments. All the components in the framework are realized in the standard of Web Service, so the other meta-schedulers or clients can interact with the components in a standard way.

We plan to complete the *Application Scheduler* to realize a more complex

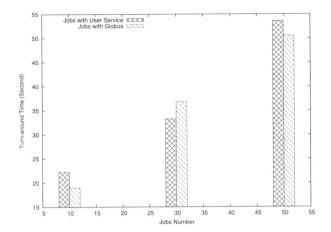

FIGURE 9.9: The comparison of submission among the User Service (Factory Service and Application Service) and Globus.

scheduling algorithm and to integrate the workflow. The *Application Scheduler* is a Web Service. The interaction between the *Application Scheduler* or between an *Application Scheduler* and the other meta-scheduler can be realized in the standard of Web service. So we would like to create a hierarchy of meta-schedulers to realize a distributed scheduling.

The rescheduling mechanism in the *Application Scheduler* should be implemented in future work. The mechanism ensures the execution of jobs, even if requested applications in some containers are removed dynamically or a container in the grid breaks down.

References

[338] Argonne National Laboratory. Getting the MPICH implementation. Available online at: `http://www-unix.mcs.anl.gov/mpi/mpich1/download.html` (Accessed September 30th, 2007).

[339] R. Bramley, K. Chiu, S. Diwan, D. Gannon, M. Govindaraju, N. Mukhi, B. Temko, and M. Yechuri. A component based services architecture for building distributed applications. In *Proceedings of HPDC, 2000*, page 51, 2000.

[340] M. C. Brown. Build grid applications based on SOA. Technical report, MCslp, 2005. Available online at: `http://www.ibm.com/developerworks/grid/library/gr-soa/` (Accessed 30th September, 2007).

[341] Condor Team. Condor user's manual. Available online at: `http://www.cs.wisc.edu/condor/manual/v6.8/2_4Road_map_Running.html` (Accessed 30th September, 2007).

[342] I. Foster. Globus toolkit version 4: Software for service-oriented systems. In *International Conference on Network and Parallel Computing (IFIP)*, volume 3779, pages 2–13. LNCS Springer-Verlag, 2005.

[343] I. Foster, C. Kesselman, J. Nick, and S. Tuecke. Grid services for distributed system integration. *IEEE Computer*, 35:37–46, 2002.

[344] I. Foster, C. Kesselman, J. Nick, and S. Tuecke. The physiology of the grid: An open grid services architecture for distributed systems integration, 2002.

[345] I. Foster, C. Kesselman, and S. Tuecke. The anatomy of the grid: Enabling scalable virtual organizations. *International Journal of High Performance Computing Applications*, 15(3):200–222, 2001.

[346] D. Gannon, R. Ananthakrishnan, S. Krishnan, M. Govindaraju, L. Ramakrishnan, and A. Slominski. Grid web services and application factories. *Computing: Making the Global Infrastructure a Reality. Fox, Berman and Hey, eds.Wiley*, 2003.

[347] Globus Team. Data Management: Key Concepts. Available online at: `http://www.globus.org/toolkit/docs/4.0/data/key/` (Accessed 30th September, 2007).

[348] Globus Team. Execution Management: Key Concepts. Available online at: `http://www.globus.org/toolkit/docs/4.0/execution/key/index.html` (Accessed 30th September, 2007).

[349] Globus Team. Globus toolkit. Available online at: `http://www.globus.org` (Accessed 30th September, 2007).

[350] Globus Team. GT 4.0 Security: Key Concepts. Available online at: `http://www.globus.org/toolkit/docs/4.0/security/key-index.html` (Accessed 30th September, 2007).

[351] Globus Team. Gt 4.0 ws gram: Job description schema doc. Available online at: `http://www.globus.org/toolkit/docs/4.0/execution/wsgram/schemas/gram_job_description.html` (Accessed 30th September, 2007).

[352] Globus Team. Submitting a job in java using WS GRAM. Available online at: `http://www.globus.org/toolkit/docs/4.0/execution/wsgram/WS_GRAM_Java_Scenarios.html` (Accessed 30th September, 2007).

[353] Gridlab. Grid(lab) grid application toolkit, 2004. Available online at: `http://www.gridlab.org/WorkPackages/wp-1` (Accessed 30th September, 2007).

[354] Gridlab. Gridlab products and technologies, 2005. Available online at: `http://www.gridlab.org/about.html` (Accessed 30th September, 2007).

[355] Y. Huang, I. Taylor, D. Walker, and R. Davies. Wrapping legacy codes for grid-based applications. In *Parallel and Distributed Processing Symposium, 2003. Proceedings. International*, 22–26 April 2003.

[356] P. Kacsuk, A. Goyeneche, T. Delaitre, T. Kiss, Z. Farkas, and T. Boczko. High-level grid application environment to use legacy codes as ogsa grid services. In *Grid Computing, 2004. Proceedings. Fifth IEEE/ACM International Workshop*, pages 428–435, 2004.

[357] G. Kandaswamy, L. Fang, Y. Huang, S. Shirasuna, and D. Gannon. A generic framework for building services and scientific workflows for the grid. In *The 2005 ACM/IEEE Conference on SuperComputing*, 2005.

[358] S. Krishnan, R. Bramley, M. Govindaraju, R. Indurkar, A. Slominski, D. Gannon, J. Alameda, and D. Alkaire. The xcat science portal. In *Proceedings SC2001*, page 49, New York, NY, USA, 2001. ACM Press.

[359] D. Kuebler and W. Eibach. Adapting legacy applications as web services. IBM DeveloperWorks, 2002. Available online at: `http://www-128.ibm.com/developerworks/library/ws-legacy/` (Accessed 30th September, 2007).

[360] C. Letondal. Pise: A tool to generate web interfaces for molecular biology programs, 2004. Available online at: `http://www.pasteur.fr/recherche/unites/sis/Pise` (Accessed 30th September, 2007).

[361] O. Lodygensky, G. Fedak, F. Cappello, V. Neri, M. Livny, and D. Thain. XtremWeb & Condor: sharing resources between Internet connected Condor pool. In *Cluster Computing and the Grid, 2003. Proceedings. CCGrid 2003. 3rd IEEE/ACM International Symposium*, pages 382–389, 12–15 May 2003.

[362] J. Mausolf. Grid in action: Monitor and discover grid services in an SOA/Web services environment, 2005. Available online at: http://www-128.ibm.com/developerworks/grid/library/gr-gt4mds/index.html (Accessed 30th September, 2007).

[363] L. Qi, H. Jin, I. Foster, and J. Gawor. HAND: Highly Available Dynamic Deployment Infrastructure for Globus Toolkit 4. 2006.

[364] J. M. Schopf, M. D'Arcy, N. Miller, L. Pearlman, I. Foster, and C. Kesselman. Monitoring and discovery in a Web services framework: Functionality and performance of the Globus Toolkit's MDS4. Technical report, Preprint ANL/MCS-P1248-0405, Argonne National Laboratory, Argonne, IL, 2005.

[365] V. Silva. Quick start to a GT4 remote execution client, 2006. Available online at: http://www-128.ibm.com/developerworks/grid/library/gr-wsgram/ (Accessed 30th September, 2007).

[366] B. Sotomayor. The globus toolkit 4 programmer's tutorial. Available online at: http://gdp.globus.org/gt4-tutorial/download/progtutorial-pdf_0.2.1.tar.gz (Accessed 30th September, 2007).

[367] B. Sundaram. Introducing gt4 security, 2005. Available online at: http://www-128.ibm.com/developerworks/grid/library/gr-gsi4intro/ (Accessed 30th September, 2007).

[368] B. Sundaram. WS-Notification and the Globus Toolkit 4 WS-Java Core, 2005. Available online at: http://www-128.ibm.com/developerworks/grid/library/gr-wsngt4/ (Accessed 30th September, 2007).

[369] W3C. Web Services Description Language (WSDL) 1.1. Available online at: http://www.w3.org/TR/wsdl (Accessed 30th September, 2007).

[370] W3C. Xml path language (Xpath) version 1.0, 1999. Available online at: http://www.w3.org/TR/xpath (Accessed 30th September, 2007).

[371] L. Yu and F. Magoulès. A Framework for Dynamic Deployment of Scientific Applications based on WSRF. In C. Cérin and K.-C. Li, editors, *Advances in grid and pervasive computing, Second International Conference, GPC 2007*, LNCS 4459, pages 579–589, Paris, France, May 2007. Springer-Verlag, Berlin, Heidelberg.

[372] L. Yu and F. Magoulès. Towards dynamic integration and scheduling of scientific applications. In *Proceedings of International Conference on Distributed Computing and Applications for Business, Engineering and Sciences*, pages 449–454, YiChang, Hubei, China, Aug. 2007. Hubei Science and Technology Press.

Chapter 10

Potential for engineering and scientific computations

10.1 Introduction

In the past ten years, grid computing has been heavily researched and developed. At the present, grid technology has matured enough to enable the realization of workable and commercial infrastructures, platforms and tools, ready to be used in production environments ranging from scientific to commercial application domains. Technology research has gained a dramatic development, involving major advances in areas such as resource management, mapping, scheduling, fault tolerance, I/O, visualization, performance analysis, programming languages and environments, resource discovery and semantic aspects.

This chapter aims at providing a comprehensive description of most recent research achievements in grid engineering. Three aspects are especially introduced to illustrate the progress of grid engineering: grid applications, some of the most important large scale projects, and grid services programming. In the grid applications section, first we introduce the common features of grid applications. Then several grid applications which are developed to solve some technical problems in industry and science research domains are introduced. In the grid projects section, several important large scale projects are presented. These projects hide the complexity of the grid and facilitate the use of grid for the non-expert users. Grid service programming is introduced at the end of this chapter. We want to emphasize that the knowledge of programming stateful Web services using GT4 is very important. This knowledge will allow you to understand the higher-level services of the toolkit and the functionality of grid components.

10.2 Grid applications

Grid technology enables integrating independent computational resources and information resources that are distributed in a wide area. Using these integrated resources, we can perform distributed/parallel computing with large-scale and distributed computing powers. Therefore, the problem of demand of large-scale computation can be solved with grid technologies.

There are some common features in applications that can use the grid environment effectively [379]. The application that is satisfied with those features is called GOCA (Grid Oriented Computing Application) and defined as follows.

- The task can be divided into several subtasks.

- The subtasks are independent of each other or have few dependencies for other subtasks.

- The subtasks can be executed in parallel.

- There is no or little communication between subtasks.

- Low cost for continuing the job when some nodes are removed.

- New nodes added to the system can be used by the job.

In order to facilitate the development of grid applications, a lot of grid middleware have been developed, such as *Globus Toolkit*. The Globus project has implemented and freely distributed grid middleware tools for security, resource management, information handling and data transfer. Therefore, the research of grid applications aims at the design of computing models and system structures which integrate the new technologies of the grid middleware and focus on the communication problem between each computing component. In this section, several grid applications are presented to illustrate the application model design and to analyze some technical issues concerning the use of grid-enabled technologies.

10.2.1 Multi-objective optimization problems solving

A multi-objective optimization problem (MOP) can be defined as the problem of finding a vector of decision variables which satisfies constraints and optimizes a vector function whose elements represent a set of objective functions [378]. MOPs are characterized by distinct measures of performance (the objectives) that may be (in)dependent. The multiple objectives being optimized almost always conflict. Thus "perfect" MOP solutions, where all decision variables satisfy associated constraints and the objective functions attain a global minimum, may not even exist [380]. Hence, the term "optimize" means finding a solution that hopefully contains values for all the objective functions that are acceptable to the designer [378].

The techniques that can be used to compute a set of solutions can be classified into heuristic methods and enumerative search. In recent years, heuristic methods have been widely studied. These methods do not guarantee obtaining the optimal solution, but they do provide near optimal solutions to a wide range of optimization problems. In contrast to heuristic methods, enumerative searching is a conceptually simple search strategy based on evaluating each possible solution in a finite search space, and thus it is able to find optimal solutions. By using some grid-enabled technology, enumerative methods can be practical to obtain optimal solutions for some MOPs.

The paper [378] uses the Globus Toolkit to implement a distributed enumerative search algorithm for solving multi-objective problems. The Globus Toolkit has been used to implement a distributed enumerative search algorithm for solving multi-objective problems. The goal is to gain experience with grid technologies to face more complex algorithms in the future. A benchmark composed of both constrained and unconstrained multi-objective problems has been developed. For unconstrained problems that are computationally expensive, promising results are worked out; in the case of constrained problems, a more parallel scheme is required to devote more resources for subranges of the constrained variables.

On the other hand, the algorithm, named gPAES, has been presented which is an approximation of grid-enabled technologies based on Globus to numerically improve an evolutionary algorithm. The algorithm uses Globus to execute a number of sequential PAES algorithms in parallel. Then, with respect to some given metric, the fronts are ranked and the best one is presented as a result.

10.2.2   Air quality predicting in a grid environment

Air Quality Forecasting (AQF) [374] is a recent discipline that addresses important air pollution problems and attempts to provide a basis for dealing with them. With the increasing maturity of air quality models, air quality forecasting services are beginning to be established. Some efforts have been made to build services, with the goal of providing timely, reliable forecasts of air quality for several regions in the US. An AQF application is created, tested and deployed and a suitable development and deployment environment is established. The AQF application makes intensive use of sophisticated numerical tools, requires high compute power for the numerical simulation of meteorological and chemical processes, and entails the transfer, storage and analysis of a huge amount of observational and simulation data.

EZ-Grid is an ongoing project that aims at making it easier and more efficient for application scientists to use grids. EZ-Grid is a lightweight, freely available implementation of a Web-based portal which provides easy access to grid functionalities. The software is very small and exhibits minimal external software dependencies, while providing a convenient interface to all functionalities of the Globus toolkit, including security, resource information, data management and job submission services.

In order to support the goal of producing reliable, timely and accurate air quality results using resources across the EZ-Grid, more work is needed. A computational grid environment must enable the specification of the complete job including the interactions between its various components; it must allow for the automated retrieval of global weather data and subsequent initiation of preprocessing; it must start the weather model once the initial data set is ready; and it must be able to launch other executables when the corresponding input data has been produced, according to the application cycle previously described. Thus, the requirements for a grid-enabled AQF are summarized in three points: workflow requirements, grid metascheduling requirements and grid security requirements.

1. In the case of workflow requirements, the Karajan software is integrated to provide workflow support in the grid environment [374]. Karajan is open source and can easily be modified to suit AQF needs. Karajan supports sequential and parallel execution containers that allow subtasks to be executed in sequence or concurrently, as desired. Yet, one major weakness of Karajan in supporting AQF is the lack of support for metascheduling and it submits the tasks of a workflow job only one by one. Thus some of the Karajan workflow components are extended to integrate with the metascheduler.

2. For the requirements of grid scheduling, first, Karajan is integrated into the metascheduler and Karajan's workflow descriptions are extended to support the global scheduling. Second, time constraints are taken into account. The local scheduler's ability is used to perform advance reservation and backfilling in order to reduce or avoid waiting time in local queues. Third, large file transfers are considered separate tasks in the workflow and the file transfer's responsibility is devolved to the metascheduler.

3. In the case of grid security requirements, a centralized CoSign authentication server is used for portal authentication. *CoSign* is an open source project at the University of Michigan to provide a Web-based authentication system.

10.2.3 Peer-to-peer media streaming systems

In the past few years, two new approaches to distributed computing have emerged, both claiming to address the problem of organizing large scale computational societies: peer-to-peer (P2P) and grid computing. At the present, the grid and P2P communities have more in common and peer resources accessed through P2P applications could be an important resource within the grid computing infrastructure.

Peer-to-Peer (P2P) networking technology has gained tremendous attention from both the academy and industry. In a P2P system, peers communicate directly with each other for the sharing and exchange of data as well as other resources such as storage and CPU capacity. Each peer acts both as a client who consumes resources from other peers, and also as a server who provides service for others. P2P systems can benefit from their following characteristics: adaptation, self-organization, load-balancing, fault-tolerance, availability through massive replication, and the ability to pool together and harness large amounts of resources [376].

A simple and straightforward way of P2P streaming implementation is to use the technique of application-layer multicast (ALM). With ALM, all peer nodes are self-organized into a logical overlay tree over the existing IP network and the streaming data are distributed along the overlay tree. From the view of network topology, current systems can be classified into three categories approximately: tree-based topology, forest-based (multi-tree) topology, and mesh topology. Various P2P media streaming systems have been proposed and developed recently. Even in China, now there are more than a dozen P2P streaming applications deployed in the Internet.

10.3   Grid projects

Computational grids promise to facilitate the sharing and integration of global resources; however, there are very few real users of grid technologies. This is partly due to the newness of grid concepts, but also because the existing infrastructure software and services are not yet mature, varied or extensive enough to provide a fully functional environment. In this context, several grid projects have been developed to provide the ability of integrating computing resources, scheduling user jobs, achieving user interfaces (grid portals) and facilitating user application developments. These projects hide the complexity of the grid and make it easy to use for non-expert users.

10.3.1   GridLab project

The GridLab is a Pan-European distributed infrastructure which consists of heterogeneous machines from various academic and research institutions. It has been established as a result of collaboration of all GridLab participants and partners in order to provide a real robust grid environment.

All GridLab technologies fit into the GridLab architecture (Figure10.1) which defines a cleanly layered environment. On the highest layer there is GAT (application oriented high level API to complex and dynamic grid environments) and GridSphere (Grid-Portal development framework). The middleware layer covers the whole range of grid capabilities as required by applications, users and administrators, such as: GRMS (Grid Resource Management and Brokering Service), Data Access and Management (Grid Ser-

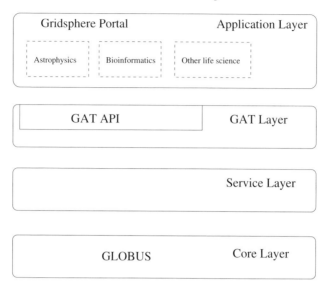

FIGURE 10.1: The GridLab architecture.

vices for data management and access), GAS (Grid Authorization Service), iGrid (GridLab Information Services), Delphoi (Grid Network Monitoring & Performance Prediction Service), Mercury (Grid Monitoring Infrastructure), Visualization (Grid Data and Visualization Services), Mobile Services (Grid Services supporting wireless technologies). GridLab technologies help real end-users to develop and run their grid-enabled applications.

The principle objective of the GridLab project is to allow the easy integration of applications with emerging grid technologies. GridLab aims to provide an environment that allows application developers to use the grid without having to understand, or even being aware of, the underlying technologies. The GAT effectively shields the application developers from the current, ever-changing grid world by providing an application-friendly interface that contains the functionality required by applications [373].

10.3.2 EU DataGrid

The EU DataGrid project (EDG) has as its aim to develop a large-scale research testbed for grid computing. The project is in its final phase and a large-scale testbed has been up and running continuously since the beginning of 2002. Three application domains are using this testbed to explore the potential grid computing has for their production environments: particle physics, earth observation, and biomedics. The EDG testbed, spanning some 20 major sites all over Europe as well as sites in the US and Asia, offers over 10,000 CPUs and 15 TB of storage to its more than 350 users; it is one of the largest grid infrastructures in the world.

The EDG Grid has a multi-layered architecture and the architecture schema is shown in Figure 10.2. The different layers from bottom to top are: the fabric layer, the underlying grid services, the data grid services and the grid application layer. At the top of the whole system, the application (e.g., Biology Application) executes an application request, submitting a grid job or requesting a file through the interfaces to the Workload Management System. The Workload Management System implements an architecture for distributed scheduling and resource management in a grid environment. It provides to the grid users a set of tools to submit their jobs, have them executed on the distributed Computing Elements, get information about their status, retrieve their output, and allow them to access grid resources in an optimal way. The Data Management System will make it possible to securely access massive amounts of data in a universal global namespace, to move and replicate data at high speed from one geographical site to another, and to manage synchronization of distributed replicas of files or databases. The Monitoring Services implement a complete infrastructure to enable end-user and administrator access to status and error information in the grid environment. The EDG collaboration has developed a complete set of tools for the management of PC farms (fabrics), in order to make the installation and configuration of the various nodes automatic and easy for the site managers managing a testbed site, and for the control of jobs on the Worker Nodes in the fabric.

10.3.3   ShanghaiGrid

As a quick response to this worldwide technical tide, a city grid project to enhance the digitalizing of a city, going by the name of ShanghaiGrid, is kicked off at the end of 2003 by the Science and Technology Commission of Shanghai municipality. The nearest goal of ShanghaiGrid is going to connect all supercomputers in this metropolis together to form a sharing environment for massive storage and grid computing [377].

The architecture of ShanghaiGrid comprises four layers, including: infrastructure layer, system software layer, supporting services layer and application layer. The infrastructure layer owns two sub-layers: hardware infrastructure sub-layer and network infrastructure sub-layer. The system software layer also owns two sub-layers: operating system sub-layer and supporting services sub-layer. In fact, all infrastructures could be regarded as the carriers of both fabric standards and connectivity protocols; system software must provide the functionalities of both the connectivity layer and resources layer (e.g., device driver, etc.); supporting services could be regarded as wrappers of local resource services and global collective services. Finally, ShanghaiGrid is a network production produced by orchestrating all these components smoothly, transparently and in a hierarchical way.

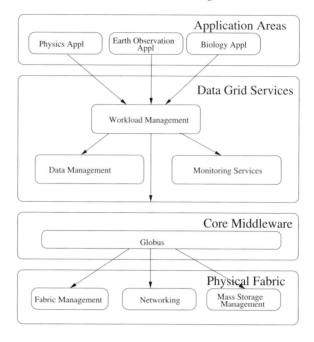

FIGURE 10.2: The EU DataGrid architecture.

10.4   Grid service programming

Currently, grid applications increasingly extend from scientific computing to commercial fields, and the Globus Toolkit (GT) has become the de facto standard of grid technologies. The latest release, the Web services-based GT4, provides significant improvements over previous releases in terms of robustness, performance, usability, documentation, standards compliance, and functionality [375]. A wide range of enabling software is included in GT4 to support the development of components that implement Web services interfaces. GT4 deals with message handling, resource management, and security, thus allowing developers to focus their attention on implementing application logic. GT4 also packages additional GT4-specific components to provide GT4 Web services containers for deploying and managing services written in Java, C, and Python. The services implemented can be divided into two types: WSRF services and non-WS services. Most of these services are implemented on top of WSRF; some services which are not implemented on top of WSRF are called the non-WS services. In fact, the WSRF implementation is a very important part of the toolkit since nearly everything else is built on top of it [366]. At this point, the knowledge of programming stateful Web services using GT4 is very important as it will allow you to progress toward using the

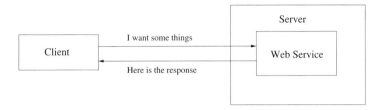

FIGURE 10.3: Web services.

higher-level services of the toolkit and to understand the functionality of grid components.

10.4.1 A short introduction to Web services and WSRF

According to the definition of W3C, a Web service is a software system designed to support inter-operable machine-to-machine interaction over a network. Even though Web services rely heavily on existing Web technologies (such as HTTP), don't mistake this with publishing something on a website. Information on a website is intended for humans. Information that is available through a Web service will always be accessed by software, never directly by a human. The clients (programs that want to access the information) contact the Web service and send a service request asking for the information. The server would return a service response. Figure 10.3 shows how a Web service works.

Web services are usually stateless. This means that the Web service can't keep state from one invocation to another. The fact that Web services don't keep state information is not necessarily a bad thing. There are plenty of applications that have no need whatsoever for statefulness. However, grid applications do generally require statefulness. So, the WSRF approach is integrated in the GT4 to enable Web services to keep state information. Giving Web services the ability to keep state information while still keeping them stateless seems like a complex problem. WSRF provides a very simple solution: simply keep the Web service and the state information completely separate. Instead of putting the state in the Web service, the state is kept in a separate entity called a resource, which will store all the state information. Each resource will have a unique key, so whenever we want a stateful interaction with a Web service we simply have to instruct the Web service to use a particular resource.

10.4.2 Java WS core programming

We can write our first stateful Web service in five simple steps.

1. Define the service's interface. This is done with WSDL.

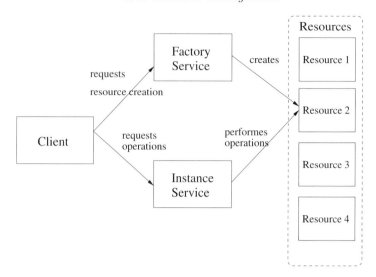

FIGURE 10.4: Multiple resource factory pattern.

2. Implement the service. This is done with Java.

3. Define the deployment parameters. This is done with WSDD and JNDI.

4. Compile everything and generate a GAR file. This is done with Ant.

5. Deploy service. This is also done with a GT4 tool.

According to the implementation of resources, we can divide these Web services into three types: single-resource, singleton resources and multiple resources. The single-resource is the simplest way to implement a stateful Web service. The service and the resource are implemented in the same class. The singleton resource splits up the implementation to *resource*, *home* and *service*, using a separate class for the service and the resource. When dealing with multiple resources, the WSRF specs recommend that we follow the factory/instance pattern, having one service in charge of creating the resources ("the factory service") and another one to actually access the information contained in the resources ("the instance service"). Figure 10.4 illustrates the multiple resource factory pattern.

The state information in the service is stored inside a resource and, more specifically, in *resource properties*. The WSRF specification, WS-ResourceProperties, has defined a set of standard PortTypes we can use to interact with a service's *resource properties*.

GetResourceProperty This PortType allows us to access the value of any resource property given its QName. This PortType provides a general way of accessing *resource properties* without the need of an individual get operation for each *resource property*.

GetMultipleResourceProperties This PortType allows us to access the value of several resource properties at once, given each of their QNames.

SetResourceProperties This PortType allows us to request one or several modifications on a service's *resource properties.*

QueryResourceProperties This PortType allows us to perform complex queries on the *resource property* document. Currently, the query language used is XPath.

For the issue of resource lifecycle management, two solutions are offered by the WS-ResourceLifetime specification: Immediate destruction and Scheduled destruction. Immediate destruction is the simplest type of lifecycle management. It allows us to request that a resource be destroyed immediately by invoking a *destroy* operation in the instance service. Scheduled destruction is a more elaborate form of resource lifecycle management, as it allows us to specify exactly when we want the resource to be destroyed.

GT4 currently supports some notification mechanisms. It allows us to effortlessly expose a resource property as a topic, triggering a notification each time the value of the *resource property* changes.

10.4.3 GT4 Security

In the industry and business domains, security issues must be taken into account. Adding security to a service does not affect the service interface. The Globus Toolkit 4 allows us to overcome the security challenges posed by grid applications through the Grid Security Infrastructure (or GSI). GSI is composed of a set of command-line tools to manage certificates, and a set of Java classes to easily integrate security into our web services. GSI offers programmers the following features:

- Transport-level and message-level security

- Authentication through X.509 digital certificates

- Several authorization schemes

- Credential delegation and single sign-on

- Different levels of security: container, service, and resource

GSI allows us to enable security at two levels: the transport level or the message level. If transport-level security is used, then the complete communication (all the information exchanged between the client and the server) would be encrypted. If we use message-level security, then only the content of the SOAP message is encrypted, while the rest of the SOAP message is left unencrypted. GSI supports three authentication methods: X.509 certificates, username and password, and anonymous authentication. GSI supports authorization in both the server-side and the client-side. Several authorization mechanisms are already included with the toolkit. In the server side, the server has six possible authorization modes: None, Self, Gridmap, Identity authorization, Host authorization and SAML Callout authorization. For the Client-side authorization, we have four authorization modes: None, Self, Identity authorization and Host. This allows the client to figure out when it will allow a service to be invoked.

10.5   Concluding remarks

Grid technologies have been widely used in industry and business domains. In this context, some of the main conceptions of grid engineering are presented. For the grid applications, we emphasize that the research of grid application should focus on the computing model and system structure design because of the development of much grid middleware which deals with security, resource management, information handling and data transfer issues in a grid environment. Then the GridLab project, EU DataGrid and ShanghaiGrid are presented as large-scale grid projects that show the generic architecture of large-scale grid system and development experiences. At the end, grid service programming is introduced. The Java WS core programming and GT4 security are two aspects presented in this section.

References

[373] G. Allen, K. Davis, K. Dolkas, N. Doulamis, T. Goodale, T. Kielmann, A. Merzky, J. Nabrzyski, J. Pukacki, T. Radke, M. Russell, E. Seidel, J. Shalf, and I. Taylor. Enabling applications on the grid-a GridLab overview. *High Performance Computing Applications*, 2003.

[374] B. M. Chapman, P. Raghunath, B. Sundaram, and Y. Yan. Predicting air quality in a production-quality grid environment. Technical report, Department of Computer Science, University of Houston, 2005.

[375] I. Foster. Globus toolkit version 4: Software for service-oriented systems. In *IFIP International Conference on Network and Parallel Computing*, LNCS 3779, pages 2–13. Springer-Verlag, 2005.

[376] W. Gao, L. Huo, and Q. Fu. Recent advances in peer-to-peer media streaming systems. *China Comminications*, 3(5):52–57, 2006.

[377] M. Li, H. Liu, F. Tang, F. Hong, C. Jiang, W. Tong, A. Zhou, Y. Gui, H. Zhu, and S. Jiang. Shanghaigrid in action: the first stage projects towards digital city and city grid. *International Journal of Grid and Utility Computing*, 1(1):22–31, 2005.

[378] F. Luna, A. Nebro, and E. Alba. Observations in using grid-enabled technologies for solving multi-objective optimization problems. *Parallel Computing*, 32:377–393, 2006.

[379] Y. Tanimura, T. Hiroyasu, M. Miki, and K. Aoi. The system for evolutionary computing on the computational grid. In *Parallel and Distributed Computing and Systems (PDCS 2002)*, pages 56–65, 2002.

[380] D. A. van Veldhuizen and G. B. Lamont. Multiobjective evolutionary algorithms: Analyzing the state-of-the-art. *Evolutionary Computation*, 8(2):125–147, 2000.

Chapter 11

Conclusions

Grid computing is analogous to the power grid in the way that computing resources will be provided in the same way as gas and electricity are provided to us now. Grid computing has evolved from metacomputing environments, such as I-WAY, that support wide-area high-performance computing to grid middleware, such as Globus toolkit, which introduces more inter-operable solutions. The current trend of grid developments is moving toward a more service oriented approach that exposes the grid protocols using Web services standards (e.g., WSDL, SOAP). This continuing evolution allows grid systems to be built in an inter-operable and flexible way and to be capable of running a wide range of applications.

11.1 Summary

11.1.1 Data management

Scientific research is now data intensive and continues to grow in size and complexity resulting in large collaborations between experimental sites and laboratories world-wide. Today information technology must cope with an increasing amount of data, which continues to increase rapidly each year. Data grids are emerging as a rather new research area, which aim to provide various services for the sharing and collaborative use of data. While the grid concepts of job scheduling and resource allocation have been widely studied, for example a variety of job scheduler and resource manager (e.g., Condor, PBS) have been developed for job and resource management in the mid-1980's, data management still has not been sufficiently addressed to fulfill the increasing data requirements of the scientific community.

Data management on grids is a challenging task due to several factors including heterogeneity at all system levels, and performance requirements associated with access, manipulation, and analysis of large amounts of data. To be efficient, data movement needs to be carefully managed between storage resources. However, the fact that in existing systems operations on data resources are embedded in the computation introduces new optimization problems. This coupling of computation with data movement causes the computation execution to delay and becomes critical for system performance as data requirements grow in size.

Grid technology has evolved to meet the challenges in terms of support for data-intensive applications as the volume and scale of data requirements for these applications increase. As a result, the main grid activities today in data-intensive computing, including major data grid projects on a worldwide scale, enforce the research for resolving the problem of large data requirements, which is vital for projects on the frontiers of science and engineering, such as high energy physics, climate modeling, earth observation, bioinformatics, and astronomy. In order to effectively provide solutions for data management in grid environments, various issues need to be considered, such as data namespace organization, a mechanism for transparent access to data resources, and efficient data transfer. Finally, an overview of existing solutions for managing data in grid environments is provided.

An overview of P2P systems is presented, underlying the characteristics of them. This overview includes certain unstructured, structured, and hybrid systems including routing algorithms for data lookup in each type of system. A table of characteristics, which summarizes the shortcomings and improvements of these systems, is also provided.

GRAVY is a grid-enabled virtual file system that enables the interoperability between heterogeneous file systems in the grid. This virtual file system integrates underlying heterogeneous file systems into a unified location-transparent file system of the grid, and provides to applications and users a uniform global view and a uniform access through standard APIs and interfaces. This approach is thus validated by a prototype implemented in Java which shows that the way users access data is simplified and that data transfers between heterogeneous file systems can be automated. This feature allows GRAVY to integrate with high-level schedulers for handling data transfer jobs.

11.1.2   Execution management

Along with the deployment of more and more heterogeneous clusters, the problem of requiring middleware to leverage existing IT infrastructure to optimize compute resources and manage data and computing workloads has emerged. Grid computing has become an increasingly popular solution to optimize resource allocation and to integrate variable computing resources in highly charged IT environments. Thus new scheduling algorithms and strategies must be researched to take into account the characteristic issues of grid. In a grid environment, the scheduling problem is to schedule a stream of applications from different users to a set of computing resources to maximize system utilization.

Eleven static heuristics and two types of dynamic heuristics are described. Then the key components of grid scheduling are presented, such as the service discovery, resource information, and grid scheduling architecture. As a specific case of application scheduling, data-intensive application scheduling is then presented. Finally, fault-tolerant technologies are discussed to deal properly with system failures and to assure the functionality of the entire grid system.

Grid workflow is increasingly used to compose complex applications in a grid environment. Workflows can be distinguished by the method mathematics which describes a workflow. There are three principal representations to present a workflow: Linear Workflow, Acyclic Graph Workflow and Cyclic Graphs Workflow. Then workflow management systems and workflow specification languages which are used to define and describe the operations and dependencies of the Workflow components are presented. As the key factors to improve the performance of workflow applications, the workflow scheduling and rescheduling theories are described and researched. Then portal projects are introduced and we point out that the portal is an important component to reduce the workflow composition time for the non-expert users. Finally, a use case, LIGO data grid infrastructure, is presented to illustrate the utilization of grid workflow.

The convergence of semantic technologies and grid computing provides many advantages. The integration of semantic technologies into Web service raises the level of description such as their capabilities and task achieving character. Thus this integration provides support in service recognition, service configuration, service comparison and automated composition. The key technologies for semantic service description have been heavily studied. But a pressing need is to develop standards and methods to describe the knowledge services themselves, and to facilitate the composition of services into larger aggregates and to negotiate workflows.

A framework that achieves the dynamic deployment of scientific applications into grid environment has been described. This framework addresses dynamic applications deployment. An Application Scheduler has been integrated in the framework, which can realize simple job scheduling and select the best grid resource to submit jobs for the users. The local administrator can dynamically make some applications available or unavailable on the grid resource without stopping the execution of the Globus Toolkit Java Web Services container. The performance of the framework has been evaluated by some experiments. All the components in the framework are realized in the standard of Web Service, so the other meta-schedulers or clients can interact with the components in a standard way.

Grid technologies have been widely used in industry and business domains. In this context, some of the main concepts of grid engineering are presented. For grid applications, we emphasize that the research should focus on the computing model and system structure design because of the development of a lot of grid middleware that deals with security, resource management, information handling and data transfer issues in a grid environment. Then the GridLab project, EU DataGrid and ShanghaiGrid are presented as large-scale grid projects which show the generic architecture of large-scale grid system and development experiences. At the end, grid service programming is introduced. The Java WS core programming and GT4 Security are two aspects to be presented.

11.2   Future for grid computing

Grid computing continues to evolve in both data management and execution management of the grid community. In the data management, we address the problem of data management in the Data Grid environment by proposing a service-oriented framework that supports explicit control of data movement scheduling and replication via file system interface. And for execution management, scheduling algorithms and strategies are presented and a framework for dynamic deployment of scientific applications is formulated. Besides these contributions, this research has raised many interesting questions and issues, that deserve further research.

Scheduling policies for data movement. Traditionally, grid workload schedulers have not taken into account data location when deciding grid site for job execution. There still remains much improvement in scheduling policies for transfers to achieve the best data-access and performance characteristics. Different scheduling policies can be studied in associating with other services, such as networking service, replication management service, job scheduler. In order to achieve the best performance, these services should work together collaboratively and in harmony. Better scheduling policies can be achieved in considering the possible ways of interaction for co-allocation of computational, storage, and network resources.

Network profiling. The properties of networks play an important role in data movement scheduling decisions. In our current implementation, the data movement scheduler does not take the changing network quality into account at execution time. A network-aware scheduler for data movement is a promising research direction in which the scheduler is aware of network conditions and able to adapt to the varying environment to achieve acceptable and predictable performance.

Integrating grid services. An important research direction is to integrate our data management service into a global grid service environment, i.e., WSRF service container. The fact that our service has been developed based on concepts and technologies from OGSA, i.e., using WSRF specification, would facilitate its integration with other WSRF-based grid services to form more sophisticated services. By following a service-oriented approach, we have put emphasis on virtualization as every resources in the system including computation resources, storages resources, networks, databases are modeled as service.

One possible solution is to integrate grid service in the portlets and using a portlet-based portal. A portlet is a web component that generates fragments-pieces of markup (e.g., HTML, XML) adhering to certain specifications (e.g., JSR-168, WSRP). Each portlet service is compounded by one or more grid service. The GridSphere portal framework is one of the candidates for service integration as it is an open-source portlet based web portal.

Deploying the middleware in a larger environment. Due to security issues, we are not able to deploy the middleware across different institutes in large scale. A natural continuation of our work would be to deploy the middleware in a larger environment in order to evaluate our approach in more realistic settings.

Implementing data-intensive scheduling. In a computational grid environment, data-intensive applications normally demand large data transfers which are costly operations in general. Therefore, taking them into account is mandatory to achieve efficient scheduling of data-intensive applications on grids. The framework for dynamic deployment of scientific applications does not take into account the bandwidth influence for scheduling and fault tolerance. Since computing resources in the grid are normally connected by wide area network links (WAN), the bandwidth limitation is an issue that must be considered when running data-intensive applications on such environments.

Several approaches have been proposed for data-intensive applications scheduling. The "bandwidth-centric" approach uses a tree structure to model a grid system, and considers the optimal solution: tasks should be allocated to nodes in order of fastest communication time. M. Faerman and his colleagues proposed a mathematical equation to evaluate the performance based on the application's communication and computational needs. The task is assigned to the resource with the best performance. Xsufferage, a modification of the Sufferage scheduling algorithm, is also a nice solution which uses the best and second-best cluster-level MCT to calculate a task's sufferage value, and assigns the task with the highest cluster-level sufferage value to the host that achieves the earliest MCT within the cluster.

Integrating semantic technologies. The convergence of semantic technologies and grid computing provides lots of advantages. The integration of semantic technologies into Web service raises the level of description such as their capabilities and task-achieving character. In the proposed framework and scheduling model, all the components are grid services which are extensions of Web services. Thus semantic technologies should be integrated to provide support in service recognition, service configuration, service comparison and automated composition.

Managing grid workflow. Grid Workflow is increasingly used to compose complex applications in a grid environment. The integration of semantic technologies into grid services provides standards and methods to describe the services knowledge, and thus facilitates the automated composition of services into larger aggregates and negotiate workflows. In order to support these future tendencies, portal technologies should be researched and integrated to discover available grid services, extract services semantic descriptions and reduce the workflow composition time for the non-expert users.

Glossary

A

Authentication The verification and validation process of the identity of a user, device, or some other computing entity, in order to allow access to resources in a system. Authentication merely ensures that the individual is who he or she claims to be, but says nothing about the access rights of the individual.

Authorization The process of granting or denying access to a resource in a system.

B

Batch job Shell scripts with control attributes.

BPEL4WS The merger of two other workflow specification languages, IBM's Web Services Flow Language (WSFL) and Microsoft's XLANG. BPEL4WS defines a model and a grammar for describing the behavior of a business process and the interactions between the process and its partners.

Business-to-Business (B2B) A type of integration technology performing business processes between trading partners.

C

Common Gateway Interface (CGI) The original technique by which a web server runs a program to dynamically create the HTML pages and to return it to the visitor's web browser.

Common Object Request Broker Architecture (CORBA) Set of industry standards published by OMG that define a distributed model for object application systems.

Component A software object encapsulating certain functionality or a set of functionalities. A component is accessed through one or more clearly defined interfaces.

Condor The goal of the Condor project is to develop, implement, deploy, and evaluate mechanisms and policies that support high throughput computing (HTC) on large collections of distributively owned computing resources.

Condor-G is the job management part of Condor. Condor-G lets the user submit the jobs into a queue, to get detailed log files of the life cycle of the jobs, to manage the input and output files, along with everything else expected from a job queuing system.

Conseil Européen pour la Recherche Nucléaire (CERN) A research laboratory with headquarters located in Geneva (Switzerland), and funded by many different countries. While most work deals with nuclear physics, CERN is known for Tim Berners-Lee's pioneering work in developing the World Wide Web portion of the Internet.

D

DAGMan A set of C libraries which allow for the user to schedule programs based on dependencies. DAGMan is part of the Condor project and extends the Condor Job Scheduler to handle intra-job dependencies.

DAML-S A DAML+OIL ontology for describing the properties and capabilities of Web services.

DAML+OIL A more recent proposal for an ontology representation language that has emerged under DARPA's Agent Markup Language (DAML) initiative along with input from leading members of the OIL consortium.

Data grid A grid infrastructure providing transparent data access to all nodes in the system through a single virtual namespace, without requiring any modifications to the client's applications. Data grid provides applications and users with a uniform interface to access data resources located across multiple locations, heterogeneous platforms and file systems, and under multiple administrative domains.

Data-intensive applications Applications that execute over a computational grid and demand large data transfers.

Distributed Component Object Model (DCOM) Microsoft's technology for distributed objects. DCOM is based on Component Object Model (COM), Microsoft's component software architecture, which defines the object interfaces. DCOM defines the remote procedure call that allows the objects to be run remotely over the network.

Distributed hash table (DHT) Class of decentralized, distributed systems and algorithms developed to provide the efficient location of data items in a very large and dynamic distributed system without relying on any centralized infrastructure. A DHT applies the principle of a hash table. A data item has an identifier. This identifier is sent to a hash function, which generates with high probability a unique key in the same virtual space. This pair of values (*identifier,key*) is completely one way, in the sense that having a similar hash value does not guarantee that the items are similar.

E

Enabling Grids and EScience in Europe (EGEE) A project integrating national, regional and thematic grid efforts to create a seamless European grid infrastructure devoted to the support of European research.

Endpoint reference (EPR) A WS-addressing construct that identifies a message destination. It consists of a Uniform Resource Identifier (URI), message reference parameters and information concerning the resource to be used.

Enterprise grids A scenario of commercial interest in which the available IT resources within a company are better exploited and the administrative overhead is lowered by the employment of grid technologies.

Exabyte (EB) Unit of storage. 1 exabyte = 10^{60} bytes.

F

Fault-tolerant algorithms Algorithms designed to deal properly with failures during the execution of a distributed algorithm.

Fault-tolerant techniques Techniques achieved to recover or to replace the failed process, in order to ensure the performance of the entire system.

File Transfer Protocol (FTP)   File exchange method that uses the Internet TCP/IP protocols to upload and download files across the Internet.

G

gLite   Next-generation middleware for grid computing. Issued from the EGEE project, gLite provides a bleeding-edge, best-of-breed framework for building grid applications tapping into the power of distributed computing and storage resources across the Internet.

Global Grid Forum (GGF)   A community-initiated forum of thousands of individuals from industries and universities of users, developers, and vendors leading the global standardization efforts of grid computing. GGF's primary objectives are to promote and to support the development, deployment, and implementation of grid technologies and applications via the creation and documentation of "best practices"-technical specifications, user experiences, and implementation guidelines.

Global grids   All kinds of resources, from single desktop machines to large-scale HPC machines, which are connected through a global grid network.

Globus alliance   A research and development project focused on enabling the application of grid concepts to develop fundamental technologies needed for building grid systems. Read more at http://www.globus.org.

Globus toolkit   An open source software toolkit used for building grids. It is being developed by the Globus alliance and many others all over the world. Read more at http://www.globus.org/toolkit.

GrADS(Grid Analysis and Display System)   An interactive desktop tool that is used for easy access, manipulation, and visualization of earth science data.

Graphical user interface (GUI)   A mechanism for interacting directly with a computing device using graphical display capabilities (such as menus, widgets, icons, and controls) to make computer applications easier to use.

Grid computing   The virtualization of distributed computing and data resources such as processing, network bandwidth and storage capacity to create a single system image, granting users and applications seamless access to vast IT capabilities. Just as an

Internet user views a unified instance of content via the Web, a grid user essentially sees a single, large virtual computer.

Grid Resource Allocation and Management (GRAM)   A Globus project that produces technologies that enable users to locate, submit, monitor and cancel remote jobs on grid-based compute resources. GRAM enables remote execution management in contexts for which reliable operation, stateful monitoring, credential management and file staging are important.

Grid resource brokers (GRBs)   Consumers in the economic model. Consumers interact with their own brokers for managing and scheduling their computations on the grid.

Grid scheduling   In a grid environment, scheduling a stream of applications from different users to a set of computing resources to maximize system utilization.

Grid service   1. (deprecated) In OGSI, a service that implements the Grid-Service PortType. 2. (informal) In its more general use, a grid service is a Web service that is designed to operate in a grid environment, and meets the requirements of the grid(s) in which it participates.

Grid service providers (GSPs)   Producers in the economic model for managing resource allocation in grid computing environments.

GridFlow   A grid workflow management system developed at the University of Warwick.

GridFTP   Grid version of the File Transport Protocol for moving large datasets between storage elements within a grid environment. Globus toolkit provides an implementation of GridFTP.

GridWay   A light-weight meta-scheduler that performs job execution management and resource brokering. It allows unattended, reliable, and efficient execution of jobs, array jobs, or complex jobs on heterogeneous, dynamic and loosely-coupled grids.

Grimoires (Grid RegIstry with Metadata Oriented Interface)   A registry for the myGrid project and the OMII grid software release (www.omii.ac.uk).

GSFL   An XML-based language that allows grid services workflow creation in the OGSA framework.

H

Heterogeneous computing (HC) The coordinated use of different types of machines, networks, and interfaces to maximize their combined performance and/or cost-effectiveness.

High performance computing grids A scenario in which different computing sites (e.g., scientific research labs) collaborate for joint research.

Hypertext Markup Language (HTML) A markup language designed for the creation of Web pages and other information viewable in a browser.

Hypertext Transfer Protocol (HTTP) The underlying communication protocol used by the World Wide Web. The protocol defines how messages are formatted and transmitted and what actions Web servers and clients (e.g., browsers) should take in response to various commands.

I

ICENI (Imperial College e-Science Networked Infrastructure) An integrated grid middleware designed to support a variety of e-science activities. These range from the exposure of resources as services, a component programming model, a scheduling framework, and the ability to visualise data from and steer applications during execution.

Information service Service that provides the accurate, up-to-date information on the structure and state of available resources in a grid environment.

J

Job A user-defined task that is scheduled to be carried out by an execution subsystem.

K

Karajan A workflow system which provides a workflow specification language and an execution engine, being developed within the Java CoG Kit.

Kepler Based on the Ptolemy II system for heterogeneous, concurrent modeling and design. With Kepler's intuitive GUI, Kepler can be used by workflow engineers and end users to design, model, execute, and reuse scientific workflows.

L

LIGO (Laser Interferometer Gravitational Wave Observatory) An ambitious effort to detect gravitational waves produced by violent events in the universe, such as the collision of two black holes, or the explosion of supernovae.

LSF   Software for managing and accelerating batch workload processing for compute- and data-intensive applications. With Platform LSF, you can intelligently schedule and guarantee completion of batch workloads across a distributed, virtualized IT environment regardless of operating system.

M

Makespan   Assume a set of jobs to be mapped into a set of machines. Completion Time can be defined as the machine availability time plus the execution time of each task on a machine in the machine set. The maximum value of Completion Time is known as the makespan.

Meta-scheduler   Enables large-scale, reliable and efficient sharing of computing resources (clusters, computing farms, servers, supercomputers, ...), managed by different LRM (Local Resource Management) systems, such as PBS, SGE, LSF, Condor, ..., within a single organization (enterprise grid) or scattered across several administrative domains (partner or supply-chain grid).

Metascheduling schemes   The hierarchy of the meta-scheduler and computing resources and the role which the meta-scheduler plays in the job scheduling.

Monitoring and Discovery System (MDS)   The information services component of the Globus toolkit providing information about the available resources on the grid and their status.

MPI   A library specification for message-passing, proposed as a standard by a broadly based committee of vendors, implementors, and users.

N

NAREGI   Started as a five-year project in 2003 as one of the major Japanese national IT projects currently being conducted. The primary objective of NAREGI is the development of the grid middleware for seamless federation of heterogeneous resources.

National science foundation (NSF)   An independent agency of the US government created in 1950 to promote the progress of science; to advance national health, prosperity, and welfare; and to secure the national defense.

Network weather service (NWS)   A distributed system that periodically monitors and dynamically forecasts the performance various network and computational resources over a given time interval.

Nimrod-G   A grid aware version of Nimrod. It takes advantage of features supported in the Globus toolkit such as automatic discovery of allowed resources.

Non-transient (persistent) sevices   A service that outlives its clients. A Web service is non-transient. It does not have the concept of service creation and destruction.

O

Object Management Group (OMG)   A consortium founded in 1989 by eleven companies (including Hewlett-Packard, IBM, Sun Microsystems, Apple Computer, American Airlines and Data General) originally aimed at setting standards for distributed object-oriented systems, and now focused on setting standards in object oriented programming as well as system modeling.

Open Grid Service Architecture (OGSA)   Standards published by Globus alliance that represent an evolution toward grid services architecture based on Web services concepts and technologies.

Open Grid Service Infrastructure (OGSI)   A GGF specification that defines the common interfaces and behaviors of a grid service. OGSI is deprecated in favor of WSRF and WS-N.

Organization for the Advancement of Structured Information Standards (OASIS)   An open international consortium that drives the development, convergence, and adoption of e-business standards. OASIS released Web services standards in different disciplines such as Web Services Resources Framework (WSRF), Web Services Notification (WS-Notification), and Web Services Security (WS-Security).

Overlay network   A virtual topology created on top of - and independently from - the underlying physical (typically IP) network.

P

P-GRADE portal The first grid portal that tries to solve the interoperability problem at the workflow level with great success.

PBS A flexible batch queueing system developed for NASA in the early to mid-1990s. It operates on networked, multi-platform UNIX environments.

Peer-to-peer (P2P) A network model where all participant nodes (i.e, peers) have identical responsibilities and are organized into an overlay network. Each peer takes both the role of client and server. As a client, it can consume resources offered from other peers, and, also as a server it can provide its services for others.

Pegasus A grid portal which provides an HTTP(S)-based interface that can be accessed using a standard web browser. The Pegasus grid portal is very useful in scenarios where a virtual organization (VO) wants to provide easy-to-use application submission interface to its members.

Petabyte (PB) Unit of storage. 1 petabyte = 10^{50} bytes.

PortType An interface that defines a set of operations performed by a service. Each operation contains a set of input, output, and fault messages. The order of these elements defines the message exchange pattern supported by the given operation.

Public key infrastructure (PKI) A system of digital certificates, certificate authorities (CA), and other registration authorities that verify and authenticate the validity of each party involved in a transaction.

Q

Quality of service (QoS) The QoS requirements for web services here mainly refer to the quality aspect of a web service. These may include performance, reliability, scalability, capacity, robustness, exception handling, accuracy, integrity, accessibility, availability, interoperability, security, and network-related QoS requirements.

R

R-GMA Part of gLite/EGEE, a monitoring and information-management service for distributed resources.

Remote procedure call (RPC) A request for a software event sent over a network. An application issues an RPC when it wants to use a function running on another system in the same network. An RPC request is synchronous, which means it requires an immediate response before the application can continue with its work, and it assumes software compatibility at each end of the communication. This tight coupling makes it best suited for use within a centrally-managed private network, rather than between separate organizations over the distributed Internet.

Resource Description Framework (RDF) A family of World Wide Web Consortium (W3C) specifications originally designed as a metadata model but which has come to be used as a general method of modeling information, through a variety of syntax formats.

S

Scheduling heuristics There are two type of scheduling heuristics: static and dynamic. These heuristics define schemes to assign tasks to machines (matching) and to compute the execution order of the tasks assigned to each machine (scheduling).

Semantic grid The convergence of the semantic Web and the grid. The semantic grid refers to an approach to grid computing in which information, computing resources and services are described using the semantic data model that can be processed by computer.

Semantic match engine The component that holds the advertisements, performs the core matching service, evaluates queries, and dynamically configures and selects matching advertisements.

Semantic web An extension of the current web in which information is given well-defined meaning, better enabling computers and people to work in cooperation.

Semantic web services Web services in which the level of description is raised and detailed in a way that indicates their capabilities and task-achieving character.

Semantic workflow Workflow in which automated composition techniques are used to automate the entire composition process by using AI planning or semantic technology.

Service An application component deployed on network-accessible platforms hosted by the service provider. Its interface is defined by a service-description language, such as WSDL, to be invoked by or to interact with a service consumer.

Service broker A repository that stores information on the available services and their locations. It is contacted by the service provider, which announces its services and contact information. The service broker is queried by service consumers to obtain the location of a service.

Service consumer An application that wants to use the functionality provided by a service. The service consumer sends a message to the provider and requests a certain service.

Service directories Called *Registry* in a grid environment. It implements the storage of arbitrary metadata about services that originate from both service providers and service users, and provides simple search APIs or web-based GUI to help requesters find Web services.

Service discovery The action of the service users or consumers to search Web services manually or automatically, after Web services are created and published in Web services registries such as UDDI.

Service oriented architecture (SOA) A framework for integrating business processes and supporting IT infrastructure as secure, standardized components - service - that can be reused and combined to address changing business priorities.

Service provider An application that has the ability to perform a certain functionality. It makes resources available to service consumers as independent services. A service provider is a self-contained, stateless business function that accepts one or more requests and returns one or more responses through a well-defined, standard interface.

Simple Mail Transfer Protocol (SMTP) Communications protocol used to transfer electronic mail messages efficiently from one server to another.

Simple Object Access Protocol (SOAP) An XML-based, platform independent protocol maintained by W3C that provides a simple and relatively lightweight mechanism for exchanging structured and typed information between services over the network. SOAP messages are independent of any operating system or protocol and can be transported using a variety of protocols, such as HTTP, SMTP, FTP, JMS, etc.

Single sign-on (SSO) A user or session authentication process that allows a user to provide one name and password and have credentials propagated to access multiple systems and applications.

Stateful sevices A service that can remember prior actions.

Stateless sevices A service that cannot remember prior actions.

T

Taverna   Developed by myGrid project, a UK e-Science pilot project building middleware to support exploratory, data-intensive, in silico experiments in molecular biology.

Terabyte (TB)   Unit of storage. 1 terabyte = 10^{40} bytes.

Transient sevices   A service that can be created and destroyed. Usually, they are created for specific clients and do not outlive their clients.

Triana   A graphical Problem Solving Environment (PSE), providing a user portal to enable the composition of scientific applications.

U

Uniform resource identifier (URI)   A generic term for all types of names and addresses that refer to objects on the World Wide Web. A URL is a type of URI.

Uniform resource locator (URL)   The address for a resource or site (usually a directory or file) on the Web and the convention that web browsers use for locating files and other remote services.

Universal Detection and Discovery Interface (UDDI)   A OASIS specification for definition of the way in which services are published and discovered across the network based on a platform independent, XML-based registry.

V

Virtual organization (VO)   A collaboration between multiple institutes. In grid computing, a VO is a community that shares resources.

Virtualization   A set of technologies and tools that enable the aggregation of resources to achieve a consolidated view throughtout an IT environment. Virtualization technologies provide a logical - rather than physical - view of data, computer power, storage capacity, and other resources.

W

Web Ontology Language (OWL)   is a language for defining and instantiating Web ontologies and is designed for use by applications that need to process the content of information instead of just presenting information to humans.

Web Service Discovery Language (WSDL) A standard language for defining a Web services description. It uses XML and XSD to describe the operations, the message formats, and protocol binding of the service.

Web services A family of technologies that consist of specifications, protocols, and industry-based standards that are used by heterogeneous applications to communicate, collaborate, and exchange information among themselves in a secure, reliable, and interoperable manner. It is the primary technology for enabling and realizing SOA concepts.

Web Services Addressing (WS-Addressing) A specification that defines XML elements to identify Web services endpoints and to provide end-to-end endpoint identification in messages. This enables messaging systems to support message transmission through networks that include processing nodes such as endpoint managers, firewalls, and gateways in a transport-neutral manner.

Web Services Base Faults (WS-BaseFaults) A specification that defines a base fault type which is used to return faults in a Web services message exchange.

Web Services Business Process Execution Language (WS-BPEL) An official OASIS standard for composition and coordination of Web services. WS-BPEL uses WSDL to describe the Web services that participate in a process and how the services interact with each other.

Web Services Resource Framework (WSRF) A family of specifications for accessing stateful resources using Web services. Since Web service implementations typically do not maintain state information during their interactions, their interfaces must allow for the manipulation of state - that is, data values that persist across and evolve as a result of Web service interactions.

Web Services Resource Lifetime (WS-ResourceLifetime) Mechanisms for WS-Resource destruction, including message exchanges that allow a requestor to destroy a WS-Resource, either immediately or by using a time-based scheduled resource termination mechanism.

Web Services Resource Properties (WS-ResourceProperties) Definition of a WS-Resource, and mechanisms for retrieving, changing, and deleting WS-Resource properties.

Web Services Resource Service Group (WS-ServiceGroup) An interface to heterogeneous by-reference collections of Web services.

Web Services Resource (WS-Resource) The core of WSRF specification. WS-Resource is defined as the composition of a resource and a Web service through which clients can access the state of this resource and manage its lifetime.

Web Services Security (WS-Security) A Web service specification that describes security enhancements to SOAP messaging, including message integrity, message confidentiality, and single message authentication.

Workflow management system A system that allows organizations to define and control the various activities (workflow) associated with a business process.

World Wide Web Consortium (W3C) An international consortium of companies involved with the Internet and the Web. The W3C was founded in 1994 by Tim Berners-Lee, the original architect of the World Wide Web. W3C's primary purpose is to develop open standards and protocols, such as HTML, HTTP, XML to ensure the universality of the Web. W3C is now heavily involved in the development of Web services standards, most notably SOAP and WSDL.

WS-Notification (WS-N) A family of specifications that defines a standard Web services approach to notification using a topic-based publish/subscribe pattern. It includes standard message exchanges to be implemented by notification broker service providers along with operational requirements expected of service providers and requestors that participate in brokered notifications. It defines also operations for a notification broker allowing publication of messages from entities that are not themselves service providers, and an XML model that describes topics.

WS-RenewableReferences A conventional decoration of a WS-Addressing endpoint reference with policy information needed to retrieve an updated version of an endpoint reference when it becomes invalid.

WSFL An XML language for the description of Web services compositions. WSFL was proposed by IBM.

X

XLANG A proposal by Microsoft Corporation for a language that is used to model business processes as autonomous agents.

XML A interoperable, self-describing data/content, in combination with XML schema definition language. Read more at http:

`//www.w3c.org/XML/`. The development of XML came because of perceived limitations in HTML when used as a tool for publishing complex documents on the Web.

XML-RPC   An XML-based standard for making simple remote calls across the networks using HTTP as transport and XML as encoding. It emerged in early 1998 as the ancestor of the SOAP protocol.

XML schema   XML documents defining the data types, content, structure, and allowed elements for an associated XML document.

Index